BIOTECHNOLOGY:
A BIBLIOGRAPHY WITH INDEXES

BIOTECHNOLOGY: A BIBLIOGRAPHY WITH INDEXES

EDWIN C. HEARNS (EDITOR)

Nova Science Publishers, Inc.
New York

Senior Editors: Susan Boriotti and Donna Dennis
Coordinating Editor: Tatiana Shohov
Office Manager: Annette Hellinger
Graphics: Wanda Serrano
Editorial Production: Jennifer Vogt, Matthew Kozlowski and Maya Columbus
Circulation: Ave Maria Gonzalez, Indah Becker, Raymond Davis and Vladimir Klestov
Communications and Acquisitions: Serge P. Shohov
Marketing: Cathy DeGregory

Library of Congress Cataloging-in-Publication Data
Available Upon Request

ISBN: 1-59033-276-8.

Copyright © 2002 by Nova Science Publishers, Inc.
400 Oser Ave, Suite 1600
Hauppauge, New York 11788-3619
Tele. 631-231-7269   Fax 631-231-8175
e-mail: Novascience@earthlink.net
Web Site: http://www.novapublishers.com

CONTENTS

PREFACE

It the end of the 20th century was a era of the Internet, the beginning of the 21st is said to be the age of biotechnology. The potential economic and social impacts are staggering. But what is it? Modern agricultural biotechnology refers to various scientific techniques, most notably genetic engineering, used to modify plants, animals, or microorganisms by introducing in their genetic makeup genes for specific desired traits, including genes from unrelated species For centuries people have crossbred related plant or animal species to develop useful new varieties or hybrids with desirable traits, such as better taste or increased productivity. Traditional crossbreeding, however, can be very time-consuming because it may require breeding several generations to obtain a desired trait and breed out numerous unwanted characteristics. Genetic engineering techniques allow faster development of new crop or livestock varieties, since the genes for a given trait can be readily introduced into a plant or animal species to produce a new variety incorporating that specific trait. Additionally, genetic engineering increases the range of traits available for developing new varieties by allowing genes from totally unrelated species to be incorporated into a particular plant or animal variety. The United States has more than a decade of experience in regulating bioengineered foods. About 50 varieties of bioengineered food crops have gone through the U.S. government regulatory procedures, and thousands of foods containing ingredients from these bioengineered crops are currently on the U.S. market. Biotechnology also contributes to such diverse areas waste disposal, mining, and medicine.

GENERAL BIBLIOGRAPHY

Advanced instrumentation, data interpretation, and control of biotechnological processes / edited by Jan F. Van Impe, Peter A. Vanrolleghem, and Dirk M. Iserentant. Published/Created: Dordrecht; Boston: Kluwer Academic Publishers, 1998. Related Authors: Impe, Jan F. van. Vanrolleghem, Peter A. Inserentant, Dirk. Community Action Programme for Education and Training for Technology. COMETT II Course on Advanced Instrumentation, Data Interpretation, and Control of Biotechnological Processes (1994: Ghent, Belgium) Description: xvii, 464 p.: ill.; 25 cm. ISBN: 0792348605 (hb: alk. paper) Notes: An outgrowth of the COMETT II Course on Advanced Instrumentation, Data Interpretation, and Control of Biotechnological Processes, held at Gent, Belgium, Oct. 1994. Includes bibliographical references. Subjects: Biotechnological process control. Biotechnological process monitoring. Biotechnology--Instruments. LC Classification: TP248.25.M65 A38 1998 Dewey Class No.: 660.6 21

Advances in biotechnology / editor, Ashok Pandey. Published/Created: New Delhi: Educational Publishers & Distributors, c1998. Related Authors: Pandey, Ashok. International Conference on Frontiers in Biotechnology (1997: Trivandrum, India) Description: xii, 514 p.: ill.; 25 cm. ISBN: 8187198036 Summary: Collection of papers presented at International Conference on Frontiers in Biotechnology, 1997 at Trivandrum. Notes: Includes bibliographical references. Subjects: Biotechnology--Congresses. LC Classification: TP248.14 .A375 1998 Dewey Class No.: 660.6 21

Advances in chitin and chitosan / edited by Charles J. Brine, Paul A. Sandford, John P Zikakis. Published/Created: London; New York: Elsevier Applied Science, c1992. Related Authors: Brine, Charles J. Sandford, Paul A. 1939- Zikakis, John P. International Conference on Chitin and Chitosan (5th: 1991: Princeton, N.J.) Description: xxii, 685 p.: ill.; 25 cm. ISBN: 1851668993 Notes: "Proceedings from the 5th International Conference on Chitin and Chitosan held in Princeton, New Jersey, USA, 17-20 October 1991"--Prelim. p. Includes bibliographical references and indexes. Subjects: Chitin--Biotechnology--Congresses. Chitosan--Biotechnology--Congresses. LC Classification: TP248.65.C55 A38 1992 Dewey Class

No.: 660/.6 20

American biotechnology laboratory.
Published/Created: [Fairfield, Conn.:
International Scientific
Communications, c1983- Description:
v.: ill.; 28 cm. Vols. for called also Dec.
1983- Current Frequency: Monthly
(with additional buyer's guide issue in
Feb.) , Former Frequency: Quarterly,
Dec. 1983- Bimonthly, Monthly, -
ISSN: 0749-3223 Incorrect ISSN: 0749-
323 Cancel/Invalid LCCN: sc 84001775
CODEN: ABLAEY Notes: Title from
cover. In 1988 the Mar., June, Sept., and
Dec. issues are tabloid issues called:
American biotechnology laboratory.
News edition; beginning in Feb. 1989
the tabloid news edition is published
bimonthly in addition to the regular
monthly issues. SERBIB/SERLOC
merged record Indx'd selectively by:
Bibliography of agriculture 0006-1530
Chemical abstracts 0009-2258 1983-
BIOTECHSEEK Jan. 1990- Life
sciences collection Subjects: Biology--
Research--Technological innovations--
Periodicals. Laboratories--Apparatus
and supplies--Periodicals.
Biotechnology--periodicals. Equipment
and Supplies--periodicals. Laboratories-
-periodicals. LC Classification: WMLC
93/2823 Dewey Class No.: 574 11

Analysis and optimization of systems / A.
Bensoussan, J.L. Lions (editors).
Published/Created: Berlin; New York:
Springer-Verlag, c1988. Related
Authors: Bensoussan, Alain. Lions,
Jacques Louis. Institut national de
recherche en informatique et en
automatique (France) International
Conference on Analysis and
Optimization of Systems (8th: 1988)
Description: xiv, 1175 p.: ill.; 25 cm.
ISBN: 0387192379 (U.S.: pbk.) Notes:

English and French. Contains most of
the 113 papers presented during the 8th
International Conference on Analysis
and Optimization of Systems, organized
by the Institut national de recherche en
informatique et en automatique, June
1988. Includes bibliographies. Subjects:
System analysis--Congresses.
Mathematical optimization--Congresses.
Automatic control--Congresses.
Biotechnology--Congresses. Series:
Lecture notes in control and information
sciences; 111 LC Classification: QA402
.A485 1988 Dewey Class No.: 003 19

Analysis of information elements used in the
assessment of certain products of
modern biotechnology / Environment
Directorate, Organisation for Economic
Co-operation and Development.
Published/Created: Paris: The
Organisation, 1995. Related Authors:
Organisation for Economic Co-
operation and Development
Description: 40 p.; 30 cm. Notes:
Includes bibliographical references (p.
39-40). Subjects: Biotechnology.
Industrial microbiology. Series: OECD
working papers; v. 3, no. 8 OECD
environment monographs; no. 100 LC
Classification: TP248.2 .A52 1995

Andrews biotechnology litigation reporter:
the twice monthly national journal of
litigation involving the biotechnology
industry. Published/Created: Wayne,
Pa.: Andrews Publications, Related
Authors: Andrews Publications, Inc.
Description: v.; 28 cm. Began in 1996?
Current Frequency: Semimonthly
Notes: Description based on: Sept. 30,
1996; title from caption.
SERBIB/SERLOC merged record
Subjects: Biotechnology industries--
Law and legislation--United States--
Digests. Genetic engineering--Law and

legislation--United States Digests. LC Classification: KF1893.B56 A16 Dewey Class No.: 343.73/0786606 21

Babel, Wolfgang. Dictionary of biotechnology: English/German / compiled by Wolfgang Babel, Monika Hagemann, Wolfgang Höhne. Published/Created: Amsterdam; New York: Elsevier, 1989. Related Authors: Hagemann, Monika. Höhne, Wolfgang. Description: 116 p.; 23 cm. ISBN: 0444989005: Notes: Includes index. Subjects: Biotechnology--Dictionaries. English language--Dictionaries-- German. LC Classification: TP248.16 .B32 1989 Dewey Class No.: 660/.6/0321 19

Bains, William, 1955- Biotechnology from A to Z / by William Bain; [foreword by G. Kirk Raab]. Edition Information: 2nd ed. Published/Created: Oxford; New York: Oxford University Press, 1998. Description: vii, 411 p.: ill.; 24 cm. ISBN: 0199636931 (pbk.) Notes: Includes bibliographical references (p. 393-394) and index. Subjects: Biotechnology--Dictionaries. LC Classification: TP248.16 .B33 1998 Dewey Class No.: 660.6/03 21

Bains, William, 1955- Biotechnology from A to Z / by William Bains. Published/Created: Oxford; New York: Oxford University Press, 1993. Description: x, 358 p.: ill.; 24 cm. ISBN: 0199633347 (pbk.) Notes: Includes index. Subjects: Biotechnology--Dictionaries. LC Classification: TP248.16 .B33 1993 Dewey Class No.: 660/.6/03 20

Barnum, Susan R. Biotechnology: an introduction / Susan R. Barnum. Published/Created: Belmont, CA:

Wadsworth Pub. Co., c1998. Description: xii, 225 p.: ill.; 28 cm. ISBN: 0534234364 (pbk.) Notes: Includes bibliographical references and index. Subjects: Biotechnology. LC Classification: TP248.2 .B363 1998 Dewey Class No.: 660/.6 21

Basic biotechnology / edited by Colin Ratledge and Bjørn Kristiansen. Edition Information: 2nd ed. Published/Created: Cambridge, U.K.; New York, NY: Cambridge University Press, 2001. Related Authors: Ratledge, Colin. Kristiansen, B. (Bjørn) Description: xiii 568 p.: ill.; 25 cm. ISBN: 0521770742 0521779170 (pbk.) Notes: Includes bibliographical references and index. Subjects: Biotechnology. LC Classification: TP248.2 .B367 2001 Dewey Class No.: 660.6 21

Basic biotechnology / edited by John Bu'Lock, Bjorn Kristiansen. Published/Created: London; Orlando: Academic Press, 1987. Related Authors: Bu'Lock, J. D. (John Desmond) Kristiansen, B. (Bjørn) Description: xiv, 561 p.: ill.; 23 cm. ISBN: 0121407535 (pbk.) 0121407527 (hard) Notes: Includes bibliographical references. Subjects: Biotechnology. LC Classification: TP248.2 .B367 1987 Dewey Class No.: 660/.6 20

Basic biotechnology: a student's guide / edited by Paul Präve ... [et al.; translated by B.J. Hazzard]. Published/Created: Weinheim; New York: VCH, c1987. Related Authors: Präve, Paul. Related Titles: [Handbuch der Biotechnology. English. Description: x, 344 p.: ill.; 24 cm. ISBN: 0895736462 (pbk.) Notes: Based on: Fundamentals of biotechnology. Includes bibliographies and index. Subjects: Biotechnology. LC Classification: TP248.2 .B37 1987 Dewey Class No.: 660/.6 19

BIO directory. Published/Created: Washington, DC: Biotechnology Industry Organization, Related Authors: Biotechnology Industry Organization. Institute for Biotechnology Information (North Carolina Biotechnology Center) Description: v.; 28 cm. -1993-94. Continued by: Biotechnology Industry Organization. Membership directory (DLC) 95640794 (OCoLC)32402369 ISSN: 1075-6434 Notes: Description based on: 1993-94. Vols. for <1993-94- prepared and designed by the Institute for Biotechnology Information of the North Carolina Biotechnology Center. SERBIB/SERLOC merged record Subjects: Biotechnology Industry Organization--Directories. Biotechnology industries--Directories. Biotechnology industries--United States--Directories. LC Classification: HD9999.B44 B53 Dewey Class No.: 338 12

Bio technology and the changing role of government. Published/Created: Paris: Organisation for Economic Co-operation and Development; Washington, D.C.: OECD Publications and Information Centre [distributor], 1988. Related Authors: Organisation for Economic Co-operation and Development. Canada-OECD Joint Workshop on National Policies and Priorities in Biotechnology (1987: Toronto, Ont.) Related Titles: Biotechnology and the changing role of government. Description: 125 p.: ill.; 23 cm. ISBN: 9264130721 Notes: Includes the report of the "Canada-OECD Joint Workshop on National Policies and Priorities in Biotechnology, Toronto, Canada, 7th-10th April, 1987". Includes bibliographical references. Subjects: Biotechnology--Government policy. LC Classification: TP248.2 .B3749 1988 Dewey Class No.: 338.4/76606 20

Bio/technology. Published/Created: [New York, N.Y.]: Nature Pub. Co., [c1983]-1996. Related Authors: Nature Publishing Company. Description: 14 v.: ill.; 28 cm. Vol. 1, no. 1 (Mar. 1983)-vol. 14, no. 2 (Feb. 1996). Current Frequency: Monthly Continued by: Nature biotechnology 1087-0156 (DLC) 96647227 (OCoLC)34160925 ISSN: 0733-222X Cancelled ISSN: 0275-7559 Cancel/Invalid LCCN: sn 82004803 sn 81002939 CODEN: BTCHDA Notes: Title from cover. SERBIB/SERLOC merged record Indx'd selectively by: Abstract bulletin of the Institute of Paper Chemistry 0020-3033 Chemical abstracts 0009-2258 Biological abstracts 0006-3169 -1990 Chemical industry notes 0045-639X Energy research abstracts 0160-3604 Mar. 1983- Excerpta medica BIOTECHSEEK 1990- Life sciences collection 1985- PESTDOC RINGDOC VETDOC Subjects: Biochemical engineering--Periodicals. Biotechnology--periodicals. LC Classification: TP248.3 .B557 Dewey Class No.: 660/.6/05 19

Biobased industrial products: priorities for research and commercialization / Committee on Biobased Industrial Products, Board on Biology, Commission on Life Sciences, National Research Council. Published/Created: Washington, D.C.: National Academy Press, c2000. Related Authors: National Research Council (U.S.). Committee on Biobased Industrial Products. Description: xiii, 147 p.; 23 cm. ISBN: 0309053927 (casebound) Notes:

Includes bibliographical references and index. Subjects: Biotechnology--United States--Forecasting. Biotechnology--Government policy--United States. LC Classification: TP248.185 .B535 2000 Dewey Class No.: 338.4/76606/0973 21

BioPeople. Published/Created: San Mateo, CA: BioVenture Publications, c1992- Description: v.: ill.; 28 cm. No. 1 (fall 1992)- Current Frequency: Quarterly ISSN: 1065-612X Cancel/Invalid LCCN: sn 92003518 CODEN: BOPEEW Notes: Title from cover. SERBIB/SERLOC merged record Indx'd selectively by: BioBusiness 1994-1996 Subjects: Biotechnology industries--United States--Periodicals. Biotechnology industries--United States--Biography Periodicals. LC Classification: HD9999.B443 U6134 Dewey Class No.: 338.4/766/06097305 20

Biopolitics: a feminist and ecological reader on biotechnology / edited by Vandana Shiva & Ingunn Moser. Published/Created: London; Atlantic Highlands, N.J.: Zed Books; Penang, Malaysia: Third World Network, 1995. Related Authors: Shiva, Vandana. Moser, Ingunn, 1962- Description: x, 294 p.; 23 cm. ISBN: 1856493350 (hb.) 1856493369 (pb.) Notes: Includes bibliographical references (p. 287-288) and index. Subjects: Biotechnology--Environmental aspects. Biotechnology--Moral and ethical aspects. Feminist theory. LC Classification: TP248.2 .B37495 1995 Dewey Class No.: 306.4/6 20

Bioresource utilization: the biotechnology option for Malaysia / edited by Ghazally Ismail. Published/Created: Petaling Jaya, Selangor Darul Ehsan, Malaysia: Pelanduk Publications, c1997. Related Authors: Ghazally Ismail. Sarawak. Ministry of Industrial Development. Universiti Malaysia Sarawak. Description: viii, 323 p.: ill.; 23 cm. ISBN: 9679786137 Notes: At head of Ministry of Industrial Development, Sarawak [and] Universiti Malaysia Sarawak. Includes bibliographical references. Subjects: Biotechnology industries--Malaysia. Biotechnology industries--Malaysia--Sarawak. LC Classification: HD9999.B44 B55 1997

Biotec ... directory. Published/Created: Habana, Cuba: Automated Information Services, Institute of Documentation and Scientific and Technical Information, Academy of Science of Cuba, Related Authors: Instituto de Documentación e Información Científica y Técnica (Academia de Ciencias de Cuba). Automated Information Services. National Center of Automated Data Exchange (Cuba) Description: v.; 29 cm. Current Frequency: Annual Cancel/Invalid LCCN: sn 92033050 Notes: Description based on: 1989. Vols. for 1990- issued by: National Center of Automated Data Exchange. SERBIB/SERLOC merged record Subjects: Biotechnology--Directories. Biotechnology--directories. Research--directories. LC Classification: TP248.17 .B55

Biotech / the Arthur Young High Technology Group. Published/Created: San Francisco: Arthur Young, c1986- Related Authors: Arthur Young High Technology Group. Ernst & Young High Technology Group. Burrill & Company. Description: v.: ill.; 27 cm. Vols. for 1999- called also 13th annual report- Vol. for 2000 called Millennium edition. 86- Current Frequency: Annual

ISSN: 1535-119X Cancel/Invalid LCCN: 89126037 Notes: Title from cover. Each issue has also a distinctive title. Published: San Franciso, Calif.: Arthur Young, 1986-1988; New York: M.A. Liebert, 1989-; San Francisco, CA: Ernst & Young, <1994-; San Francisco, CA: Burrill & Company, <1998; Palo Alto, CA: Ernst & Young, 1999- Vols. for 1986-1989 issued by: Arthur Young High Technology Group; 1990- by: Ernst & Young High Technology Group; <1998- by: Burrill & Company. Additional Form Avail.: Summary of report for 1999- available also online via the World Wide Web, in PDF file format. Subjects: Biotechnology industries--United States--Periodicals. LC Classification: HD9999.B443 U627 Dewey Class No.: 338.4/76606/0973 20

Biotech '95: ALI-ABA course of study materials: business, law, and regulation, November 2-3, 1995, San Francisco, California / cosponsored by California Continuing Education of the Bar. Published/Created: Philadelphia, PA (4025 Chestnut St., Philadelphia 19104-3099): American Law Institute-American Bar Association Committee on Continuing Professional Education, c1995. Related Authors: California Continuing Education of the Bar. American Law Institute-American Bar Association Committee on Continuing Professional Education. Description: xiii, 327 p.: ill., forms; 28 cm. Notes: "CA03." Includes bibliographical references. Subjects: Biotechnology industries--Law and legislation--United States. LC Classification: KF1893.B56 B545 1995 Dewey Class No.: 346.7303/8 347.30638 20

Biotech USA 1987: proceeding of the conference held in Santa Clara, California, November 1987. Published/Created: London: Online, c1987. Related Authors: Online Publications (Firm) Description: xiii, 377 p.: ill.; 28 cm. ISBN: 0863531113 Notes: "The 4th annual industry conference and exhibition"--P. 4 of cover. Includes bibliographies and index. Subjects: Biotechnology--United States--Congresses. Biotechnology industries--United States--Congresses. LC Classification: TP248.14 .B555 1987 Dewey Class No.: 338.4/766/060973 20

Biotechnologies / SRI International, Business Intelligence Program. Published/Created: Menlo Park, CA: The Program, Related Authors: Business Intelligence Program (SRI International) Description: v.; 28 cm. Current Frequency: Annual ISSN: 1058-4625 Notes: "The year in review." Description based on: 1988; title from cover. SERBIB/SERLOC merged record Subjects: Biotechnology industries--Periodicals. LC Classification: HD9999.B44 B57 Dewey Class No.: 338.4/76606/05 20

Biotechnologies in perspective: socio-economic implications for developing countries / edited by Albert Sasson and Vivien Costarini. Published/Created: Paris: Unesco, 1991. Related Authors: Sasson, Albert. Costarini, Vivien. Description: 166 p.; 24 cm. ISBN: 9231027387 Notes: "Future-oriented studies/Études prospectives"--Cover. Includes bibliographical references. Subjects: Biotechnology industries--Developing countries Congresses. LC Classification: HD9999.B443 D482

1991

Biotechnology / organized and edited by C.F. Phelps and P.H. Clarke. Published/Created: London: Biochemical Society, 1983. Related Authors: Phelps, C. F. (Charles Frederick) Clarke, P. H. (Pat H.) Biochemical Society (Great Britain) Description: ix, 247 p.: ill.; 26 cm. ISBN: 0904498158: Notes: "Held at University College London, December 1982." Includes bibliographies and index. Subjects: Biotechnology--Congresses. Series: Biochemical Society (Great Britain). Symposium. Biochemical Society Symposia; no. 48. Variant Series: Biochemical Society symposia, 0067-8694; no. 48 LC Classification: QH345 .B522 no. 48 TP248.2 Dewey Class No.: 574.1/92/08 s 620.8 19

Biotechnology and Biological Sciences Research Council (Great Britain) Annual report / Biotechnology and Biological Sciences Research Council. Published/Created: Swindon, England: BBSRC, [1995- Description: v.: ill.; 22 x 28 cm. 1994/95- Current Frequency: Annual Absorbed: Agricultural and Food Research Council (Great Britain). Annual report (1994) (OCoLC)41094686 ISSN: 1360-4791 Notes: SERBIB/SERLOC merged record Subjects: Biotechnology and Biological Sciences Research Council (Great Britain)--Periodicals. Biotechnology--Research--Great Britain--Periodicals. Life sciences--Research--Great Britain--Periodicals. Biology--Research--Great Britain--Periodicals. LC Classification: TP248.195.G7 B56b Dewey Class No.: 660.6/072041 21

Biotechnology and Biological Sciences Research Council (Great Britain) Handbook / Biotechnology and Biological Sciences Research Council. Published/Created: Swindon [England]: BBSRC, 1995- Description: v.; 30 cm. 1994/95- Current Frequency: Biennial Continues: Agricultural and Food Research Council (Great Britain). Handbook 0961-1010 (DLC)sn 91016309 (OCoLC)23033919 Incorrect ISSN: 0961-1010 Cancel/Invalid LCCN: sn 95015526 Notes: SERBIB/SERLOC merged record Subjects: Biotechnology and Biological Sciences Research Council (Great Britain)--Directories. Biotechnology--Research grants--Great Britain Periodicals. Biology--Research grants--Great Britain--Periodicals. Life scientists--Great Britain--Directories. LC Classification: TP248.195.G7 B56a Dewey Class No.: 660/.6/072041 21

Biotechnology and biosafety: a forum / cosponsored by American Association for the Advancement of Science ... [et al.]; Ismail Serageldin and Wanda Collins, editors. Published/Created: Washington, D.C.: World Bank, c1999. Related Authors: Serageldin, Ismail, 1944- Collins, Wanda W. (Wanda Williams) American Association for the Advancement of Science. International Conference on Environmentally Sustainable Development (5th: 1997: World Bank) Description: ix, 214 p.; 28 cm. ISBN: 0821342428 Notes: "Biotechnology and biosafety was a forum associated with the Fifth Annual World Bank Conference on Environmentally and Socially Sustainable Development, held at the World Bank, October 9-10, 1997." Includes bibliographical references. Subjects: Biotechnology--

Environmental aspects--Congresses. Health risk assessment--Congresses. Agricultural biotechnology--Congresses. Series: Environmentally and socially sustainable development series. Variant Series: Environmentally and socially sustainable development LC Classification: TP248.2 .B55116 1999 Dewey Class No.: 660.6 21

Biotechnology and competitive advantage: Europe's firms and the US challenge / edited by Jacqueline Senker; co-ordinated by Ronald van Vliet. Published/Created: Cheltenham, UK; Northampton, US: E. Elgar, c1998. Related Authors: Senker, Jacqueline. Description: xii, 180 p.; 24 cm. ISBN: 1858987393 Notes: Includes bibliographical references and index. Subjects: Biotechnology industries--European Union countries. Biotechnology industries--United States. Competition, International. LC Classification: HD9999.B443 E8513 1998 Dewey Class No.: 338.4/76606/094 21

Biotechnology and culture: bodies, anxieties, ethics / edited by Paul E. Brodwin. Published/Created: Bloomington: Indiana University Press, c2000. Related Authors: Brodwin, Paul. Description: vi, 296 p.; 24 cm. ISBN: 0253214289 (pbk.: alk. paper) 025333831X (cl: alk. paper) Notes: Includes bibliographical references and index. Subjects: Biotechnology--Social aspects. Series: Theories of contemporary culture; v. 25 LC Classification: TP248.2 .B55117 2000 Dewey Class No.: 303.48/3 21

Biotechnology and the consumer: a research project sponsored by the Office of Consumer Affairs of Industry Canada / edited by Bartha Maria Knoppers and Alan D. Mathios. Published/Created: Dordrecht; Boston: Kluwer Academic Publishers, c1998. Related Authors: Knoppers, Bartha Maria. Mathios, Alan D. Description: xiv, 509 p.: 1 ill.; 25 cm. ISBN: 0792355415 (alk. paper) Notes: "Partly reprinted from the Journal of consumer policy, volume, 21, no. 4, 1998." Includes bibliographical references. Subjects: Biotechnology. Consumers. Biotechnology--Canada. Biotechnology--Public opinion. Biotechnology industries. Biotechnology--Moral and ethical aspects. LC Classification: TP248.2 .B55128 1999 Dewey Class No.: 338.4/76606 21

Biotechnology annual review. Published/Created: Amsterdam; New York: Elsevier, c1995- Description: v.: ill.; 25 cm. Vol. 1 (1995)- Current Frequency: Annual ISSN: 1387-2656 CODEN: BAREFD Notes: SERBIB/SERLOC merged record Indx'd selectively by: Index medicus 0019-3879 v4, 1998- Chemical abstracts 0009-2258 Subjects: Biotechnology--Periodicals. LC Classification: TP248.13 .B573 Dewey Class No.: 660.6 21

Biotechnology Centre research report [microform]. Published/Created: New Delhi: Indian Agricultural Research Institute, 1988. Related Authors: Indian Agricultural Research Institute. Description: x, 36 p.: col. ill.; 25 cm. Notes: Master microform held by: DLC. Includes bibliographical references (p. 34-36). Microfiche. New Delhi: Library of Congress Office; Washington, D.C.: Library of Congress Photoduplication Service, 1996. 1 microfiche. Subjects: Biotechnology. LC Classification:

Microfiche 96/60172 (T)

Biotechnology in Asia: development strategies, applications, and potentials. Published/Created: Tokyo: Asian Productivity Organization, 1990. Related Authors: Asian Productivity Organization. APO Study Meeting on Biotechnology (1989: Osaka, Japan) Description: ii, 412 p.: ill.; 26 cm. ISBN: 9283320905 Notes: "Report of APO Study Meeting on Biotechnology held in Japan in September 1989"-- Verso t.p. Includes bibliographical references. Subjects: Biotechnology-- Asia--Congresses. Biotechnology industries--Asia--Congresses. LC Classification: TP248.195.A78 B56 1990 Dewey Class No.: 338.4/76606/095 20

Biotechnology in Colorado / compiled by the Department of Microbiology and Environmental Health, Colorado State University and R.D. Speer Associates; [Ralph Smith, Rhett Speer, principal investigators]. Published/Created: Fort Collins, Colo.: Colorado State University; Boulder, Colo.: R.D. Speer Associates, c1984. Related Authors: Smith, Ralph, 1940- Speer, R. D. Colorado State University. Dept. of Microbiology and Environmental Health. R.D. Speer Associates. Colorado Commission on Higher Education. Description: iv, 432 p.: ill.; 28 cm. Notes: "Assembled in response to requests from the Colorado Commission on Higher Education ... [et al.]"--P. 1. "March 1984." Subjects: Biotechnology--Research--Colorado-- Directories. Biotechnologists-- Colorado--Directories. LC Classification: TP248.17 .B56 1984 Dewey Class No.: 660/.6/0720788 19

Biotechnology in developing countries: symposium, Delft, The Netherlands, 13 and 14 October 1982 / editors, P.A. van Hemert, H.L.M. Lelieveld, J.W.M. la Rivière. Published/Created: [Delft]: Delft University Press, 1983. Related Authors: Hemert, Paulus Aloysius van. Lelieveld, H. L. M. La Rivière, J. W. M. Nederlandse Vereniging voor Microbiologie. Technical Microbiology Section. Netherlands Biotechnological Society. Description: viii, 158 p.: ill.; 24 cm. ISBN: 9062751385 (pbk.): Notes: "Organized on behalf of the Technical Microbiology Section of the Netherlands Society for Microbiology [and] the Netherlands Biotechnological Society, Section of the Royal Netherlands Chemical Society"--P. ii. Subjects: Biotechnology--Developing countries--Congresses. LC Classification: TP248.2 .B553 1983 Dewey Class No.: 660.6 19

Biotechnology in industry: selected applications and unit operations / by Rajani Joglekar ... [et al.]. Published/Created: Ann Arbor, Mich.: Ann Arbor Science, c1983. Related Authors: Joglekar, Rajani. Description: xv, 179 p.: ill.; 24 cm. ISBN: 0250406055 Notes: Includes bibliographical references and index. Subjects: Biotechnology. LC Classification: TP248.3 .B6 1983 Dewey Class No.: 660/.6 19

Biotechnology in Japan: the reference source / edited by Japan Pacific Associates. Published/Created: Bunkyo-ku, Tokyo: Science Forum; Palo Alto, CA: Distributed in U.S. by Japan Pacific Associates, c1984. Related Authors: Japan Pacific Associates. Saiensu F‾oramu Kabushiki Kaisha. Description: 322, 10 p.: ill.; 28

cm. Subjects: Biotechnology--Japan--Directories. Biotechnology industries--Japan--Directories. LC Classification: TP248.17 .B57 1984 Dewey Class No.: 338.4/766/060952 19

Biotechnology in New Zealand / compiled by D.M. Hunt ... [et al.]; [edited by G.F. Preddey]. Published/Created: Wellington: New Zealand Dept. of Scientific and Industrial Research, 1983. Related Authors: Hunt, D. M. Preddey, G. F. Description: 82 p.: ill.; 30 cm. ISBN: 047706731X (pbk.) Notes: Bibliography: p. 79-80. Subjects: Biotechnology--New Zealand. Series: DSIR discussion paper, 0110-5221; no 8 LC Classification: TP248.2 .B555 1983 Dewey Class No.: 660.6/09931 19

Biotechnology in the developing world and countries in economic transition / [edited by] G.T. Tzotzos, K.G. Skryabin. Published/Created: Wallingford: CABI Pub., c2000. Related Authors: Tzotzos, George T. Skriabin, K. G. (Konstantin Georgievich) Description: xv, 312 p.: ill.; 26 cm. ISBN: 0851993311 Notes: Includes bibliographical references. Subjects: Biotechnology--Developing countries. Biotechnology industries--Developing countries. LC Classification: TP248.195.D48 B556 2000 Dewey Class No.: 338.4/76606/091724 21

Biotechnology processes: scale-up and mixing / edited by Chester S. Ho, James Y. Oldshue. Published/Created: New York, N.Y.: American Institute of Chemical Engineers, c1987. Related Authors: Ho, Chester S., 1950- Oldshue, James Y. American Institute of Chemical Engineers. Description: vii, 267 p.: ill.; 29 cm. ISBN: 0816904103

Notes: Includes bibliographies and index. Subjects: Biochemical engineering--Congresses. Biotechnology--Methodology--Congresses. LC Classification: TP248.3 .B62 1987 Dewey Class No.: 660/.6 19

Biotechnology progress. Published/Created: New York, NY: American Institute of Chemical Engineers, c1985- Related Authors: American Institute of Chemical Engineers. American Institute of Chemical Engineers. Food, Pharmaceutical, and Bioengineering Division. American Chemical Society. Description: v.: ill.; 28 cm. Vol. 1, no. 1 (Mar. 1985)- Current Frequency: Bimonthly, 1990- Former Frequency: Quarterly, 1985-1989 ISSN: 8756-7938 Cancel/Invalid LCCN: sn 85000467 CODEN: BIPRET Notes: Title from cover. Vols. for 1985-1989 issued by the Food, Pharmaceutical, and Bioengineering Division of AIChE; 1990- cosponsored by the American Chemical Society and the American Institute of Chemical Engineers. SERBIB/SERLOC merged record Vols. for 1985-1989 include FPBE news as a special section. Earlier and later issues of FPBE news are published separately. Indexed by: Bibliography of agriculture 0006-1530 Indexed entirely by: Applied science & technology index 0003-6986 1991- Indx'd selectively by: Biological abstracts 0006-3169 1985- Chemical abstracts 0009-2258 1985- BIOTECHSEEK 1990- Additional Form Avail.: Also available via the World Wide Web; online version available to subscribers with a site license; files in HTML and PDF format. Subjects: Biotechnology--Periodicals. Genetic engineering--Periodicals. Biomedical Engineering--periodicals. Food Technology--periodicals.

Technology, Pharmaceutical--periodicals. Food industry and trade--Periodicals. Bioengineering--Periodicals. LC Classification: TP248.13 .B578 Dewey Class No.: 660/.6/05 20

Biotechnology, a Dutch perspective / edited by J.H.F. van Apeldoorn. Published/Created: Delft: Delft University Press, 1981. Related Authors: Apeldoorn, J. H. F. van. Stichtung Toekomstbeeld der Techniek. Description: x, 248 p.: ill.; 24 cm. ISBN: 9062750516 (pbk.) Notes: A study sponsored by the Netherlands Study Centre for Technology Trends. Includes bibliographical references. Subjects: Biotechnology. Series: STT publications; 30 LC Classification: TP248.3 .B59 Dewey Class No.: 660/.6 19

Biotechnology, current progress. Published/Created: Lancaster, Pa.: Technomic Pub. Co., c1991. Description: 1 v.: ill.; 27 cm. Vol. 1. Current Frequency: Annual ISSN: 1055-2162 Cancel/Invalid LCCN: sn 91001774 CODEN: BCUPE6 Notes: SERBIB/SERLOC merged record Indx'd selectively by: Biological abstracts 0006-3169 1991- Subjects: Biotechnology--Periodicals. Biotechnology--periodicals. Technology, Medical--periodicals. LC Classification: TP248.13 .B5755 Dewey Class No.: 660/.6/05 20

Biotechnology: a hope or a threat? / edited by Iftikhar Ahmed; foreword by Michael Lipton. Published/Created: New York: St. Martin's Press, 1992. Related Authors: Ahmed, Iftikhar, 1944- Description: xxvii, 275 p.: ill.; 23 cm. ISBN: 031207154X Notes: Includes bibliographical references (p. 247-261) and index. Subjects: Biotechnology--Social aspects. LC Classification: TP248.2 .B523 1992 Dewey Class No.: 303.48/3 20

Biotechnology: applications and research / edited by Paul N. Cheremisinoff, Robert P. Ouellettee, in collaboration with R.M. Bartholomew. Published/Created: Lancaster: Technomic Pub., c1985. Related Authors: Cheremisinoff, Paul N. Ouellette, Robert P., 1938- Bartholomew, R. M. (Richard M.) Description: xiv, 699 p.: ill.; 26 cm. ISBN: 0877623910 Notes: Includes bibliographies and index. Subjects: Biotechnology. LC Classification: TP248.2 .B5514 1985 Dewey Class No.: 660/.6 19

Biotechnology: commercialization and economic aspects: January 1993 - June 1996 / compiled by Scott A. Leonard and Raymond Dobert. Published/Created: Beltsville, Md.: National Agricultural Library, [1996] Related Authors: Leonard, Scott A. Dobert, Raymond. National Agricultural Library (U.S.) Description: 37 p.; 28 cm. Notes: "151 citations in English from AGRICOLA." "Updates QB 94-20." Shipping list no.: 97-0038-P. "September 1996." Includes indexes. Subjects: Biotechnology--Economic aspects--Bibliography. Genetic engineering--Economic aspects--Bibliography. Series: Quick bibliography series; 96-10. Variant Series: Quick bibliography series, 1052-5378; QB 96-10 LC Classification: Z7914.B33 B562 1996 TP248.2 Dewey Class No.: 338.4/76606 21

Biotechology Canada. Published/Created: [Willowdale, Ont.: Sentry

Communications, 1988-1989] Related Titles: [Canadian research. Description: 2 v.: ill.; 29 cm. Vol. numbering continues a publication which published periodically from Jan. 1986-Oct. 1987 as demographic issues of Canadian research; they also constitute and carry the regular vol. numbering of Canadian research. [Vol. 3, no. 1] (winter 1988)-v. 4, no. 2 (spring 1989). Current Frequency: Quarterly ISSN: 0842-0998 Incorrect ISSN: 0319-1974 Cancel/Invalid LCCN: cn 89039021 Notes: "A demographic issue of Canadian research." Also available on microfilm from Maclean Hunter Microfilm Services, Toronto. SERBIB/SERLOC merged record Merged with: Canadian research, ISSN 0319-1974; and, Laboratory times, to become: Canadian laboratory, ISSN 0848-8002. Supplement to: Canadian research 0319-1974 (DLC) 84647686 (OCoLC)2442252 (CaOONL)750349395 Subjects: Biotechnology--Canada--Periodicals. LC Classification: TP248.13 .B575 Dewey Class No.: 660/.6 19

Biotecnology: 1986 regulatory update: November 12, 1986, Radisson Mark Plaza Hotel, Alexandria, Virginia / sponsored by Center for Energy and Environmental Management. Published/Created: [Fairfax Station, Va.]: Center for Energy and Environmental Management, c1986. Related Authors: Center for Energy and Environmental Management (Fairfax Station, Va.) Description: xii, 209, 104 p.: ill.; 30 cm. Subjects: Biotechnology industries--Law and legislation--United States--Congresses. Genetic engineering--Law and legislation--United States Congresses. LC Classification: KF1893.B56 B55 1986

Dewey Class No.: 353.0085/5 19

California. Legislature. Senate. Committee on Finance, Investment, and International Trade. Implications of genetically modified organisms on international trade. Published/Created: Sacramento, CA: Senate Publications, [2000] Related Authors: California. Legislature. Senate. Subcommittee on California-European Trade Development. Description: 1 v. (various pagings): ill.; 28 cm. Notes: Cover title. "Joint informational hearing, Senate Committee on Finance, Investment, and International Trade and Subcommittee on California-European Trade and Development." Transcript of a hearing held Apr. 26, 2000, State Capitol, Sacramento. Senate Publications stock no.: 1049-S. Subjects: Crops--Genetic engineering--California. Food--Biotechnology--California. Food--Labeling--Standards--California. Crops-Genetic engineering. Food--Biotechnology. Food--Labeling--Standards. LC Classification: KFC10.3 .F56 2000

Canada. National Biotechnology Advisory Committee. Annual report / National Biotechnology Advisory Committee. Published/Created: [Ottawa]: Ministry of State, Science and Technology, c1985-c1989. Related Authors: Canada. Ministry of State, Science and Technology. Description: 3 v.; 28 cm. Report for 1984 covers Oct. 1983 to Nov. 1984. 1984-1987-1988. Current Frequency: Annual Continued by: Canada. National Biotechnology Advisory Committee. Report (DLC) 99100806 (OCoLC)31200356 ISSN: 0839-5713 Cancel/Invalid LCCN: ce 88071993 Notes: Text in English and French with French text on inverted pages. SERBIB/SERLOC merged record Subjects: Biotechnology--Canada--Periodicals. LC Classification: TP248.13 .C36a Dewey Class No.: 338.97106 19

Canada. National Biotechnology Advisory Committee. Report / National Biotechnology Advisory Committee. Added TP Rapport Published/Created: [Ottawa]: Ministry of Supply and Services Canada, c1991- Description: v.; 28 cm. 4th (1989-1990)- Current Frequency: Irregular Continues: Canada. National Biotechnology Advisory Committee. Annual report 0839-5713 (DLC) 86649446 (OCoLC)13646152 Notes: Text in English and French, with French text on inverted pages, 1989-1990-; text in English only, <1998- SERBIB/SERLOC merged record Also available in French, <1998- Also available in alternative formats, <1998- Also available online. Subjects: Biotechnology--Canada--Periodicals. LC Classification: TP248.13 .C36a Dewey Class No.: 338.97106 19

Coombs, J. Dictionary of biotechnology / J. Coombs. Published/Created: New York: Elsevier, c1986. Description: 330 p.: ill.; 25 cm. ISBN: 0444010874 044401070X (pbk.) Subjects: Biotechnology--Dictionaries. LC Classification: TP248.16 .C66 1986 Dewey Class No.: 660/.6/0321 19

Crafts-Lighty, A. (Anita), 1954- Information sources in biotechnology / A. Crafts-Lighty. Edition Information: 2nd ed. Published/Created: New York, N.Y.: Stockton Press, c1986. Description: 403 p.: ill.; 30 cm. ISBN: 0943818184 (pbk.) Notes: Includes bibliographies and index. Subjects: Biotechnology--

Bibliography. Biotechnology--Information services--Directories. LC Classification: Z7914.B33 C7 1986 TP248.2 Dewey Class No.: 660/.6/07 19

Crafts-Lighty, A. (Anita), 1954- Information sources in biotechnology / A. Crafts-Lighty. Published/Created: New York, N.Y.: Macmillan: Nature Press, 1983. Description: 306 p.; 30 cm. ISBN: 0943818044 (Nature Press: pbk.): Notes: Includes bibliographies and index. Subjects: Biotechnology--Bibliography. Biotechnology--Information services--Directories. LC Classification: Z7914.B33 C7 1983 TP248.2 Dewey Class No.: 660/.6/07 19

Current opinion in biotechnology. Published/Created: London, UK: Current Biology, c1990- Description: v.: ill.; 28 cm. Vol. 1, no. 1 (Oct. 1990)- Current Frequency: Bimonthly ISSN: 0958-1669 Cancel/Invalid LCCN: sn 90031458 CODEN: CUOBE3 Notes: Title from cover. Includes bibliography of the current world literature. SERBIB/SERLOC merged record Indexed entirely by: Index medicus 0019-3879 1995- Indx'd selectively by: Biological abstracts 0006-3169 1992- BIOTECHSEEK Oct. 1990- Chemical abstracts 0009-2258 1990- Additional Form Avail.: Also available via World Wide Web via the Science Direct web site and at OCLC FirstSearch Electronic Collections Online; Subscription required for access to abstracts and full text. Subjects: Biotechnology--Periodicals. Biotechnology--Bibliography--Periodicals. Biotechnology--bibliography. Biotechnology--periodicals. Series: Current opinion series LC Classification: TP248.13 .C87 Dewey

Class No.: 660/.6 20

Directory of biotechnology centers / North Carolina Biotechnology Center, Information Division. Published/Created: Research Triangle Park, NC: North Carolina Biotechnology Center, c1989- Related Authors: North Carolina Biotechnology Center. Information Division. Description: v.; 28 cm. 1989- Continues: Directory of states' biotechnology centers (OCoLC)21794527 Cancel/Invalid LCCN: sn 89024138 Notes: Title from cover. SERBIB/SERLOC merged record Subjects: Biotechnology--United States--Directories--Periodicals. LC Classification: TP248.17 .D53 Dewey Class No.: 660/.6/02573 20

Directory of biotechnology companies. U.S. companies, western region I, California. Published/Created: San Diego, Calif.: A. Gee Associates, Related Authors: A. Gee Associates. Description: v.; 28 cm. Current Frequency: Annual ISSN: 1075-1912 Notes: Description based on: 2nd ed. published in 1992. SERBIB/SERLOC merged record Subjects: Biotechnology industries--California--Directories. LC Classification: HD9999.B443 U633 Dewey Class No.: 338.7/6606/025794 20

E.B.C.-Symposium on Biotechnology (1983: Nutfield, Surrey) E.B.C.--Symposium on Biotechnology, Nutfield, Great Britain, November 1983 / European Brewery Convention. Published/Created: Nürnberg: Brauwelt-Verlag, 1984. Description: xvi, 294 p.: ill.; 24 cm. ISBN: 3418006515 (pbk.) Notes: Includes bibliographies. Subjects: Biotechnology--Congresses.

Series: Monograph (European Brewery Convention); 9. Variant Series: Monograph / European Brewery Convention; 9 LC Classification: TP248.2 .E18 1983 Dewey Class No.: 660/.6 19

Enari, Tor-Magnus. From beer to molecular biology: the evolution of industrial biotechnology / Tor-Magnus Enari. Published/Created: Nürnberg: Fachverlag Hans Carl, c1999. Description: 120 p.: ill.; 23 cm. ISBN: 3428007708 Notes: Includes bibliographical references (p. 120). Subjects: Biotechnology--History. Industrial microbiology--History. Biotechnology--history. Beer--history. Fermentation. LC Classification: TP248.18 .E53 1999 Dewey Class No.: 660.6/09 21

Fleschar, Manfred H. Glossary of biotechnology terms / Manfred H. Fleschar, Kimball R. Nill. Published/Created: Lancaster, PA: Technomic Pub. Co., c1993. Related Authors: Nill, Kimball R. Description: vi, 153 p.; 24 cm. ISBN: 0877629919 Subjects: Biotechnology--Dictionaries. LC Classification: TP248.16 .F54 1993 Dewey Class No.: 660/.6/03 20

Grace, Eric S. Biotechnology unzipped: promises & realities / Eric S. Grace. Published/Created: Washington, D.C.: Joseph Henry Press, c1997. Description: xv, 248 p.: ill.; 23 cm. ISBN: 0309057779 Notes: Includes bibliographical references (p. 235-237) and index. Subjects: Biotechnology--Popular works. Biotechnologie--Ouvrages de vulgarisation. LC Classification: TP248.15 .G73 1997 Dewey Class No.: 660.6 21

Gray, Lynn. The new future of biotechnology: enabling technologies and star products / Lynn Gray, Rahul Jasuja. Published/Created: Norwalk, Conn. (25 Van Zant St., Norwalk 06855): Business Communications Company, c2001. Related Authors: Jasuja, Rahul. Business Communications Co. Description: xvi, 181, [11] leaves: ill.; 28 cm. ISBN: 156965588X Notes: "January 2001." Includes bibliographical references (leaf xiii). Subjects: Biotechnology industries--United States. Pharmaceutical industry--United States. Biotechnology industries--United States--Directories. Pharmaceutical industry--United States--Directories. Market surveys--United States. Series: Business opportunity report; B-147 LC Classification: HD9999.B443 U639 2001

Guenther, Kim. Biotechnology, public perception: January 1992 - June 1994 / Kim Guenther and Raymond Dobert. Published/Created: Beltsville, Md.: National Agricultural Library, [1994] Related Authors: Dobert, Raymond. National Agricultural Library (U.S.) Description: 30 p.; 28 cm. Notes: "154 citations in English from AGRICOLA." "Updates QB 93-15." Shipping list no.: 94-0324-P. "September 1994." Includes indexes. Subjects: Biotechnology--Public opinion--Bibliography. Series: Quick bibliography series; 94-58. Variant Series: Quick bibliography series, 1052-5378; QB 94-58 LC Classification: Z7914.B33 G843 1994 TP248.2

Haaland, Perry D., 1952- Experimental design in biotechnology / Perry D. Haaland. Published/Created: New York: Marcel Dekker, c1989.

Description: xv, 259 p.: ill.; 24 cm. ISBN: 0824778812 (alk. paper) Notes: Includes bibliographical references. Subjects: Biotechnology--Experiments--Statistical methods. Biotechnology--Experiments--Graphic methods. Experimental design. Series: Statistics, textbooks, and monographs; v. 105 LC Classification: TP248.24 .H33 1989 Dewey Class No.: 660/.6/0724 20

Hodgson, J. G. (John G.), 1937- Biotechnology: changing the way nature works / John Hodgson. Published/Created: London: Cassell; New York, N.Y.: Distributed by Sterling Pub. Co., 1989. Description: 124 p.: col. ill.; 28 cm. ISBN: 0304317837: Notes: Distributor from label on t.p. verso. "An Equinox book"--T.p. verso. Subjects: Biotechnology--Popular works. LC Classification: TP248.215 .H63 1989 Dewey Class No.: 660/.6 20

Industrial biotechnology / editors, Vedpal S. Malik, Padma Sridhar; co-editors, M.C. Sharma, H. Polasa. Published/Created: New Delhi: Oxford & IBH Pub. Co., 1992. Related Authors: Malik, Vedpal S. Sridhar, Padma. Sharma, M. C. Polasa, H. Description: x, 621 p.: ill.; 25 cm. ISBN: 8120407105

International biotechnology laboratory. Published/Created: [Fairfield, Conn.: International Scientific Communications, Inc., Related Authors: International Scientific Communications, Inc. Description: v.: ill.; 28 cm. Current Frequency: Bimonthly, <1993- Former Frequency: Quarterly, <1984- ISSN: 0888-7225 Cancel/Invalid LCCN: sn 86014371 Notes: Description based on: Vol. 2, no.

4 (Dec. 1984); title from cover. SERBIB/SERLOC merged record Subjects: Biotechnology--Methodology--Periodicals. Biotechnology laboratories--Periodicals. Biotechnology--periodicals. Equipment and Supplies--catalogs. LC Classification: TP248.13 .I59 Dewey Class No.: 660/.6 19

International computer-based conference on biotechnology: a case study / editor, D.A. Balson. Published/Created: Ottawa, Canada: International Development Research Centre, c1985. Related Authors: Balson, David. Description: 108 p.; 25 cm. ISBN: 0889364451 (pbk.) Notes: Summaries in French and Spanish. Includes bibliographical references. Subjects: Biotechnology--Information services--Congresses. Teleconferencing--Congresses. Computer networks--Congresses. Computer conferencing. Series: IDRC (Series); 241e. Variant Series: IDRC; 241e LC Classification: TP248.14 .I57 1985

International developments in biotechnology and their possible impact on certain sectors of the U.S. chemical industry: report on investigation no. 332-174 under Section 332 (b) of the Tariff Act of 1930. Published/Created: Washington, D.C.: U.S. International Trade Commission, [1984] Related Authors: United States International Trade Commission. Description: xxi, 164 p.; 28 cm. Notes: Cover title. "October 1984." Includes bibliographical references. Subjects: Chemical industry--United States. Biotechnology industries. Biotechnology--Industrial applications. Market surveys. Series: USITC publication; 1589 LC Classification:

HD9651.5 .I58 1984 Dewey Class No.: 338.4/766/00973 19

International Symposium on Analytical Methods, Systems, and Strategies in Biotechnology (4th: 1992: Noordwijkerhout, Netherlands) Analytical biotechnology: proceedings of the 4th International Symposium on Analytical Methods, Systems, and Strategies in Biotechnology, Noordwijkerhout, The Netherlands, September 21-23, 1992 / edited by C. Van Dijk. Published/Created: Amsterdam; New York: Elsevier, 1993. Related Authors: Dijk, C. van. Description: 202 p.: ill.; 27 cm. ISBN: 0444816402 (acid-free paper) Notes: Papers presented at the 4th International Symposium on Analytical Methods, Systems and Strategies in Biotechology, held in Noordwijkerhout, The Netherlands, September 21-23, 1992. "Reprinted from Analytica chimica acta and Journal of biotechnology." Includes bibliographical references and index. Subjects: Biotechnology--Technique--Congresses. LC Classification: TP248.24 .I58 1992 Dewey Class No.: 660/.63 20

International Symposium on Biotechnology (1987: Bratislava, Czechoslovakia) Interbiotech '87, enzyme technologies: proceedings of the International Symposium on Biotechnology, Bratislava, Czechoslovakia, June 25-26, 1987 / edited by A. Blazej and J. Zemek. Published/Created: Amsterdam; New York: Elsevier, 1988, c1987. Related Authors: Blazej, Anton. Zemek, J. Description: 535 p.: ill.; 25 cm. ISBN: 0444989242: Notes: Includes bibliographies. Subjects: Enzymes--Industrial applications--Congresses. Biotechnology--

Congresses. Series: Progress in biotechnology; 4 LC Classification: TP248.E5 I58 1987 Dewey Class No.: 660/.63 19

Los Alamos science / Los Alamos Scientific Laboratory. Published/Created: [Los Alamos, NM]: The Laboratory, [1980?- Related Authors: Los Alamos Scientific Laboratory. Los Alamos National Laboratory. Description: v.: ill. (some col.); 28 cm. Issue for summer 1980 called also Premier issue. [Vol. 1, no. 1] (summer 1980)-v. 3, no. 3 (Fall 1982); no. 7 (winter/spring 1983)- Current Frequency: Annual, 1992- Former Frequency: Four no. a year, 1980-198 Irregular, 198 -1990 ISSN: 0273-7116 Cancel/Invalid LCCN: sn 80002602 CODEN: LASCDI Notes: Full-text CD-ROM accompanies occasional issue. Title from cover. Vols. for summer 1980-198 issued by: the Los Alamos Scientific Laboratory; by: the Los Alamos National Laboratory. SERBIB/SERLOC merged record Indx'd selectively by: Mathematical reviews 0025-5629 GeoRef 0197-7482 Chemical abstracts 0009-2258 Additional Form Avail.: Issue for 1993 distributed to depository libraries in microfiche. Also issued online. Subjects: Los Alamos Scientific Laboratory--Periodicals. Los Alamos Scientific Laboratory. Los Alamos National Laboratory. Science--Periodicals. Technology--Periodicals. Biological Sciences--periodicals. Biotechnology--periodicals. Science--periodicals. Technology--periodicals. LC Classification: Q11 .L5 Dewey Class No.: 505

Made not born: the troubling world of biotechnology / edited by Casey Walker. Published/Created: New York: Sierra Club Books, 2000. Projected Pub. Date: 1111 Related Authors: Walker, Casey. Description: p. cm. ISBN: 1578050596 (alk. paper) Notes: Includes bibliographical references and index. Subjects: Biotechnology--Popular works. Biotechnology industries--Popular works. LC Classification: TP248.215 .M33 2000 Dewey Class No.: 660.6 21

Maryland Biotechnology Institute. Maryland Biotechnology Institute: the first five years, 1985-1990. Published/Created: [College Park, Md.: Maryland Biotechnology Institute, 1990] Related Titles: First five years, 1985-1990. Description: 40 p.: ill.; 28 cm. Notes: Cover title. Bibliography: p. 27-38. Subjects: Maryland Biotechnology Institute. Biotechnology. Biotechnology--Research--Maryland. High technology. LC Classification: TP248.19.M3 M37 1990 Dewey Class No.: 660/.6/0720752 20

National Seminar on "Potentials and Prospects of Biotechnology in Developing Countries" (1989: School of Biotechnology, Jawaharlal Nehru Technological University) Proceedings of National Seminar on "Potentials and Prospects of Biotechnology in Developing Countries", 6.2.1989 to 8.2.1989. Published/Created: Hyderabad: School of Biotechnology, Jawaharlal Nehru Technological University, [1989] Related Authors: Jawaharlal Nehru Technological University. School of Biotechnology. Description: 82 p.: ill.; 30 cm. Notes: Cover title. Includes bibliographical references. Subjects: Biotechnology--Congresses. Biotechnology--Developing countries--Congresses. LC Classification: TP248.14 .N36 1989

Dewey Class No.: 338.4/76606/091724 20

National Symposium on Biotechnology (1st: 1982: Chandigarh, India) Biotechnology: proceedings of the First National Symposium on Biotechnology, at Chandigarh, March 13-15, 1982 / editor, S.C. Jain. Published/Created: Chandigarh: Dept. of Chemical Engineering and Technology, Panjab University, [1983] Related Authors: Jain, S. C. (Suresh C.), 1926- Panjab University. Dept. of Chemical Engineering and Technology. Description: 359 p., [6] p. of plates: ill.; 26 cm. Notes: Sponsor: Department of Chemical Engineering and Technology, Panjab University. Includes bibliographical references and index. Subjects: Biotechnology--Congresses. LC Classification: TP248.2 .N38 1982 Dewey Class No.: 660/.6 19

Ouellette, Robert P., 1938- Applications of biotechnology / Robert P. Ouellette, Paul N. Cheremisinoff. Published/Created: Lancaster, Pa.: Technomic Pub. Co., c1985. Related Authors: Cheremisinoff, Paul N. Description: xii, 247 p.: ill.; 23 cm. ISBN: 0877624380 (pbk.) Notes: Includes bibliographies and index. Subjects: Biotechnology. Biotechnology industries. LC Classification: TP248.2 .O94 1985 Dewey Class No.: 660/.6 19

Ouellette, Robert P., 1938- Essentials of biotechnology / Robert P. Ouellette, Paul N. Cheremisinoff. Published/Created: Lancaster, Pa.: Technomic Pub. Co., c1985. Related Authors: Cheremisinoff, Paul N. Description: ix, 226 p.: ill.; 23 cm. ISBN: 0877624372 (pbk.) Notes: Includes bibliographies and index.

Subjects: Biotechnology. LC Classification: TP248.2 .O944 1985 Dewey Class No.: 660/.6 19

Rapp, Barbara A. (Barbara Ann) Biotechnology information sources: North and South America / compiled and edited by Barbara A. Rapp. Published/Created: Medford, N.J.: Learned Information, Inc., 1994. Related Authors: International Council of Scientific and Technical Information. Description: viii, 144 p.; 28 cm. ISBN: 0938734814 Notes: "Published for the International Council for Scientific and Technical Information." Includes bibliographical references (p. [115]-117) and index. Subjects: Biotechnology--North America--Information services Directories. Biotechnology--South America--Information services Directories. Biotechnology--North America--Databases--Directories. Biotechnology--South America--Databases--Directories. Biotechnology--bibliography. Biotechnology--directories. Information Systems. Information Services--North America--directories. Information Services--South America--directories. Databases, Factual. Databases, Bibliographic. LC Classification: TP248.17 .R37 1994 Dewey Class No.: 025.06/6606 21

Rimmington, Anthony. Technology and transition: a survey of biotechnology in Russia, Ukraine, and the Baltic States / Anthony Rimmington with Rod Greenshields. Published/Created: Westport, Conn.: Quorum Books, 1992. Related Authors: Greenshields, Rod, 1933- Description: ix, 227 p.: 1 map; 24 cm. ISBN: 089930804X (alk. paper) Notes: Includes bibliographical references (p. [186]-218) and index.

Subjects: Biotechnology--Russia (Federation). Biotechnology--Ukraine. Biotechnology--Baltic States. LC Classification: TP248.195.R8 R56 1992 Dewey Class No.: 338.4/76606/0947 20

Seidman, Lisa A. Basic laboratory methods for biotechnology: textbook and laboratory reference / Lisa A. Seidman, Cynthia J. Moore. Published/Created: Upper Saddle River, N.J.: Prentice Hall, c2000. Related Authors: Moore, Cynthia J. Description: vi, 751 p., 4 p. of plates: ill. (some col.); 28 cm. ISBN: 0137955359 (spiral bound) Notes: Includes bibliographical references and index. Subjects: Biotechnology--Laboratory manuals. LC Classification: TP248.24 .S45 2000 Dewey Class No.: 660.6/078 21

Stableford, Brian M. The Cassandra Complex / Brian Stableford. Edition Information: 1st ed. Published/Created: New York: Tor, 2001. Description: 319 p.; 22 cm. ISBN: 0312877730 (alk. paper) Notes: "A Tom Doherty Associates book." Subjects: Twenty-first century--Fiction. Forensic scientists--Fiction. Missing persons--Fiction. Biotechnology--Fiction. Genre/Form: Science fiction. LC Classification: PR6069.T17 C375 2001 Dewey Class No.: 823/.914 21

State-by-state biotecnology directory: centers, companies, and contacts / prepared by the Biotechnology Information Division of the North Carolina Biotechnology Center. Published/Created: Washington, D.C.: Bureau of National Affairs, c1990. Related Authors: North Carolina Biotechnology Center. Information Division. Description: xiv, 163 p.; 28 cm. ISBN: 1558711732: Notes:

Includes index. Subjects: Biotechnology--United States--Directories. LC Classification: TP248.17 .S72 1990 Dewey Class No.: 660/.6/02573 20

Sukatsch, Dieter A. Biotechnology: a handbook of practical formulae / Dieter A. Sukatsch, Alexander Dziengel. Published/Created: Burnt Mill, Harlow, Essex, England: Longman Scientific & Technical; New York: Wiley, 1987. Related Authors: Dziengel, Alexander. Description: 160 p.; 20 cm. ISBN: 0470207299 (Wiley: pbk.) Notes: Translation of: Formelsammlung Biotechnologie. Includes index. Bibliography: p. 142. Subjects: Biotechnology--Formulae--Handbooks, manuals, etc. Biotechnology--Nomenclature--Handbooks, manuals, etc. LC Classification: TP248.162 .S8513 1987 Dewey Class No.: 660/.6 19

The Biotech business: financial outlook and analysis. Published/Created: Washington, D.C.: Special Projects Unit of the Bureau of National Affairs, c1989. Related Authors: Bureau of National Affairs (Washington, D.C.) Description: v, 134 p.: ill.; 28 cm. ISBN: 1558711562: Notes: Cover title. Subjects: Biotechnology industries--United States--Finance. Series: The BNA special report series on biotechnology; special report #3 LC Classification: HD9999.B443 U614 1989 Dewey Class No.: 338.4/36606/0973 20

The biotech revolution: analysis of future technology & markets. Published/Created: New York, NY: Technical Insights, c1998. Related Authors: Technical Insights, Inc.

Description: xxii, 197 p.; 28 cm. ISBN: 0471329703 Subjects: Biotechnology industries--Forecasting. Biotechnology--Industrial applications. Series: Technical insights; R-236. LC Classification: HD9999.B442 B5643 1998 Dewey Class No.: 338.4/76606 21

The biotech revolution: analysis of future technology & markets. Published/Created: Englewood, N.J.: Wiley, c1998. Description: xxii, 197 p.; 28 cm. Subjects: Biotechnology. Biotechnology industries. Series: Technical insights LC Classification: TP248.2 .B376 1998 Dewey Class No.: 338.4/76606 21

The Biotechnology directory. Published/Created: New York, N.Y.: Nature Press, 1985- Related Authors: Coombs, J. Description: v.: ill.; 30 cm. Vols. for <1986- called also <3rd ed.-1985- Current Frequency: Annual Continues in part: International biotechnology directory (DLC) 85645263 (OCoLC)10960249 Absorbed: International biotechnology directory (London, England) 0265-3877 (DLC)sn 86017409 (OCoLC)12774709 ISSN: 1059-7352 Cancel/Invalid LCCN: sn 85019879 sn 89044069 Notes: "Products, companies, research and organizations." Published: New York, N.Y.: Stockton Press, 1986- Editor: 1985- J. Coombs. SERBIB/SERLOC merged record Issued also in a British edition with International biotechnology directory. Subjects: Biotechnology industries--Directories. Biochemical engineering--Directories. Bionics--Directories. Biomedical Engineering--directories. Genetic Engineering--directories. Technology, Medical--directories. LC Classification: TP248.3 .I56 Dewey

Class No.: 660/.6/025 19

Trends in biotechnology. Published/Created: [Amsterdam, Netherlands: Elsevier Science Publishers, c1983- Description: v.: ill.; 30 cm. Vol. 1, no. 1 also called inaugural issue. Vol. 1, no. 1 (Mar./Apr. 1983)- Current Frequency: Monthly Former Frequency: Bimonthly ISSN: 0167-7799 Incorrect ISSN: 0167-9430 0166-9430 Cancel/Invalid LCCN: sn 84010498 CODEN: TRBIDM Notes: Title from caption. Imprint varies: Cambridge, UK: Elsevier Publications Cambridge, Edition statement supplied. SERBIB/SERLOC merged record Also issued in an annual compilation called: Reference ed. Indx'd selectively by: Index medicus 0019-3879 v16n5,May 1998- Chemical abstracts 0009-2258 1983- Excerpta medica BIOTECHSEEK Dec. 1990- Life sciences collection Additional Form Avail.: Available also by subscription via the World Wide Web. Subjects: Biotechnology--Periodicals. Biochemistry--methods--periodicals. Biomedical Engineering--trends--periodicals. Technology--trends--periodicals. Biochemical engineering--Periodicals. Genetic engineering--Periodicals. Industrial microbiology--Periodicals. LC Classification: TP248.13 .T74 Dewey Class No.: 660/.6/05 19

U.S.-Israel Research Conference on Advances in Applied Biotechnology (1990: Haifa, Israel) Biotechnology: bridging research and applications: proceedings of the U.S.-Israel Research Conference on Advances in Applied Biotechnology, June 24-30, 1990, Haifa, Israel / edited by Daphne Kamely, Ananda M. Chakrabarty, Steven E. Kornguth. Published/Created: Boston:

Kluwer Academic Publishers, c1991.
Related Authors: Kamely, Daphne.
Chakrabarty, Ananda M., 1938-
Kornguth, Steven E. Description: xiv,
459 p.: ill.; 25 cm. ISBN: 0792311442
(acid-free paper) Notes: Includes
bibliographical references and index.
Subjects: Biotechnology. LC
Classification: TP248.2 .B5515 1991
Dewey Class No.: 660/.6 20

Wang, Daniel I-chyau, 1936-
Biotechnology: status and perspectives /
Daniel I.C. Wang. Published/Created:
New York, N.Y.: American Institute of
Chemical Engineers, 1988. Related
Authors: American Institute of
Chemical Engineers. Meeting (1986:
Miami Beach, Fla.) Description: 22 p.:
ill.; 28 cm. ISBN: 0816904375 Notes:
Cover title. Institute lecturer, 1986
Annual Meeting of the American
Institute of Chemical Engineers, Miami
Beach, Florida, November 2-7, 1986.
Bibliography: p. 22. Subjects:
Biotechnology. Series: AIChE
monograph series; no. 18, vol. 84 LC
Classification: TP248.2 .W36 1988
Dewey Class No.: 660/.6 19

Ward, Owen P., 1947- Bioprocessing /
Owen P. Ward. Edition Information:
U.S.A. ed. Published/Created: New
York: Van Nostrand Reinhold, 1991.
Description: x, 198 p.: ill.; 24 cm.
ISBN: 0442314396 Notes: Includes
bibliographical references and index.
Subjects: Biotechnology--Technique.
LC Classification: TP248.24 .W37 1991
Dewey Class No.: 660/.6 20

MEDICINE

Advanced healing technologies.
Published/Created: [Wellesley Hills, Mass.]: Business Technology Research, c1988. Related Authors: Business Technology Research (Firm) Description: 185 p.: ill.; 29 cm. Subjects: Wound treatment equipment industry. Biotechnology industries. Market surveys. LC Classification: HD9995.W682 A38 1988 Dewey Class No.: 338.4/761714 20

Advanced medical technology / project director, Edward D. Hester.
Published/Created: Cleveland, Ohio: The Freedonia Group, c1992. Related Authors: Hester, Edward. Freedonia Group. Description: vi, 183 leaves; 28 cm. Notes: "March 1992." Subjects: Medical instruments and apparatus industry--United States. Biotechnology industries--United States. Market surveys--United States. Series: Industry study (Freedonia Group); 406. Variant Series: Industry study; 406 LC Classification: HD9994.U52 A38 1992

Alberta biotechnology and pharmaceutical directory. Published/Created: Edmonton, Alta.: Alberta Economic Development and Tourism, Technology Development Branch, [1995?] Related Authors: Alberta. Alberta Economic Development and Tourism. Technology Development Branch. Description: 28 p.: col. ill., map; 28 cm. Notes: Cover title. Subjects: Biotechnology industries--Alberta--Directories. Pharmaceutical industry--Alberta--Directories. LC Classification: HD9999.B443 C262 1995

Alliance alert. Medical/health.
Published/Created: Needham, MA: Venture Economics, c1990- Related Authors: Venture Economics, Inc. Description: v.; 27 cm. Vol. 1, issue 1 (Apr. 1990)- Current Frequency: Quarterly ISSN: 1053-0649 Incorrect ISSN: 1050-0367 Cancel/Invalid LCCN: sn 90003467 Notes: Title from cover. SERBIB/SERLOC merged record Companion publication to: Alliance alert. Electronics/computer hardware/industrial automation; Alliance alert. Software/information services; Alliance alert. Communications; and: Alliance alert. Chemicals/materials/agriculture. Alliance alert. Electronics/computer hardware/industrial automation 1050-0367 (DLC) 90660963 Alliance alert. Software/information services 1053-0665 (DLC) 90660964 Alliance alert. Communications 1053-0657 (DLC)sn 90003472 (OCoLC)22445560 Alliance

alert. Chemicals/materials/agriculture 1053-0673 (DLC) 91656082 Subjects: Medical instruments and apparatus industry--United States Directories. Pharmaceutical industry--United States--Directories. Biotechnology industries--United States--Directories. Consolidation and merger of corporations--United States Directories. Foreign licensing agreements--United States--Directories. Corporations, Foreign--United States--Directories. Foreign licensing agreements--United States--Directories. Joint ventures--United States--Directories. International business enterprises--United States Directories. LC Classification: HD9994.U5 A45 Dewey Class No.: 338.7/613621/02573 20

Amalgamations: fusing technology and culture / edited by Susanne Lundin and Lynn Åkesson. Published/Created: Lund: Nordic Academic Press, c1999. Related Authors: Lundin, Susanne. Åkesson, Lynn. Description: 137 p.: 22 cm. ISBN: 9189116070 Notes: Includes bibliographical references. Subjects: Technology and civilization. Biotechnology. Biomedical Technology. Anthropology, Cultural. Ethnology. Information Systems. LC Classification: HM221 .A54 1999

Artificial cells, blood substitutes, and immobilization biotechnology. Published/Created: Monticello, NY: M. Dekker, [1994- Related Authors: International Society for Artificial Cells and Immobilization Biotechnology. Description: v.: ill.; 23 cm. Vol. 22, no. 1- Current Frequency: Five no. a year Continues: Biomaterials, artificial cells, and immobilization biotechnology 1055-7172 (DLC) 92649899 (OCoLC)23274995 ISSN: 1073-1199

Cancel/Invalid LCCN: sn 93005601 CODEN: ABSBE4 Notes: Title from cover. Issued by International Society for Artificial Cells and Immobilization Biotechnology. SERBIB/SERLOC merged record Indexed entirely by: Index medicus 0019-3879 1994- Additional Form Avail.: Also available to subscribers of OCLC FirstSearch Electronic Collections Online and Dekker via the World Wide Web. Subjects: Artificial cells--Periodicals. Immobilized cells--Periodicals. Immobilized ligands (Biochemistry)--Periodicals. Blood substitutes--Periodicals. Biotechnology--Periodicals. Biocompatible Materials--periodicals. Biological Products--periodicals. Biotechnology--periodicals. Blood Substitutes--periodicals. LC Classification: R856.A1 B56 Dewey Class No.: 610.28 20

Biomarkets: forecasts and analyses of 40 opportunities. Published/Created: Englewood/Fort Lee, NJ: Technical Insights, c1995. Related Authors: Technical Insights, Inc. Description: xiii, 157 leaves; 30 cm. ISBN: 156217018X Subjects: Biotechnology industries--Forecasting. Pharmaceutical industry--Forecasting. Biotechnology industries--Directories. Pharmaceutical industry--Directories. Market surveys. LC Classification: HD9999.B442 B547 1995

Biomaterials regulating cell function and tissue development: symposium held April 13-14, 1998, San Francisco, California, U.S.A. / editors, Robert C. Thomson ... [et al.]. Published/Created: Warrendale, Pa.: Materials Research Society, c1998. Related Authors: Thomson, Robert C. Description: vii, 119 p.: ill.; 24 cm. ISBN: 155889436X

Notes: Includes bibliographical references and indexes. Subjects: Biomedical materials--Congresses. Animal cell biotechnology--Congresses. Polymers in medicine--Congresses. Biocompatible Materials--congresses. Biomedical Engineering--congresses. Cell Communication--physiology--congresses. Bone Regeneration--congresses. Series: Materials Research Society symposia proceedings; v. 530. Variant Series: Materials Research Society symposium proceedings; v. 530 LC Classification: R857.M3 B5734 1998 Dewey Class No.: 610/.28 21

Biomaterials, artificial cells, and immobilization biotechnology: official journal of the International Society for Artificial Cells and Immobilization Biotechnology. Published/Created: New York, N.Y.: Marcel Dekker, 1991-1993. Related Authors: International Society for Artificial Cells and Immobilization Biotechnology. Description: 3 v.: ill.; 23 cm. Vol. 19, no. 1-v. 21, no. 5. Current Frequency: 5 no. a year, 1992-1993 Former Frequency: Four no. a year, 1991 Continues: Biomaterials, artificial cells, and artificial organs 0890-5533 (DLC) 87644604 (OCoLC)14264306 Continued by: Artificial cells, blood substitutes, and immobilization biotechnology 1073-1199 (DLC) 94641007 (OCoLC)29333477 ISSN: 1055-7172 Cancel/Invalid LCCN: sn 91000588 CODEN: BACBEU Notes: Title from cover. SERBIB/SERLOC merged record Indexed entirely by: Excerpta medica 1991- Index medicus 0019-3879 1991- Indx'd selectively by: BioBusiness 1991- Biological abstracts 0006-3169 1991- Chemical abstracts 0009-2258 1991- Computer & control abstracts 0036-8113 1991- Electrical & electronics abstracts 0036-8105 1991-

Energy research abstracts 0160-3604 1991- Engineering index bioengineering abstracts 0736-6213 1991- Engineering index monthly (1984) 0742-1974 1991- Life sciences collection 1991- Metals abstracts 0026-0924 1991- Physics abstracts 0036-8091 1991- Subjects: Biomedical materials--Periodicals. Artificial cells--Periodicals. Artificial organs--Periodicals. Immobilized cells--Periodicals. Biocompatible Materials--periodicals. Biological Products--periodicals. Biotechnology--periodicals. LC Classification: R856.A1 B56 Dewey Class No.: 610/.28 20

Biomaterials: products, trends, and markets. Published/Created: Tustin, CA, U.S.A. (17722 Irvine Blvd., Tustin 92680): Biomedical Business International, c1985. Related Authors: Biomedical Business International, Inc. Description: 1 v. (various pagings): ill.; 29 cm. Notes: "August 1985." Subjects: Medical instruments and apparatus industry. Pharmaceutical industry. Sanitary supply industry. Biotechnology industries. Market surveys. Series: Report (Biomedical Business International, Inc.); #7061. Variant Series: Report / Biomedical Business International; #7061 LC Classification: HD9994.A2 B56 1985 Dewey Class No.: 381/.456606/0973 19

Biomedical importance of marine organisms / edited by Daphne, G. Fautin. Published/Created: San Francisco, Calif.: California Academy of Sciences, 1988. Related Authors: Fautin, Daphne Gail. California Academy of Sciences. Symposium (3rd: 1987: San Francisco, Calif.) Description: viii, 159 p.: ill.; 28 cm. ISBN: 0940228203 (pbk.) Notes: "Third biennial Symposium of the California Academy of Sciences"--P.

vii. Includes bibliographical references. Subjects: Marine pharmacology--Congresses. Marine biotechnology--Congresses. Marine organisms--Congresses. Series: Memoirs of the California Academy of Sciences; no. 13 Memoirs of the California Academy of Sciences (1988); no. 13. LC Classification: RS160.7 .B57 1988

Bio-medical materials and engineering. Published/Created: New York: Pergamon Press, 1991- Description: v.: ill.; 26 cm. Vol. 1, no. 1- Current Frequency: Quarterly ISSN: 0959-2989 Cancel/Invalid LCCN: sn 91001382 CODEN: BMENEO Notes: Title from cover. SERBIB/SERLOC merged record Indexed entirely by: Index medicus 0019-3879 1991- Indx'd selectively by: Chemical abstracts 0009-2258 Additional Form Avail.: Also available via World Wide Web; OCLC FirstSearch Electronic Collections Online; Subscription required for access to abstracts and full text. Subjects: Biomedical materials--Periodicals. Bioengineering--Periodicals Biocompatible Materials--periodicals. Biomedical Engineering--periodicals. Biotechnology--periodicals. LC Classification: R857.M3 B48 Dewey Class No.: 610/.28 20

Biomedical materials--drug delivery, implants, and tissue engineering: symposium held November 30-December 1, 1998, Boston, Massachusetts, U.S.A. / editors, Thomas Neenan, Michele Marcolongo, Robert F. Valentini. Published/Created: Warrendale, Pa.: Materials Research Society, c1999. Related Authors: Neenan, Thomas. Marcolongo, Michele. Valentini, Robert F. Materials Research Society. Description: xiii, 376 p.: ill.; 24 cm. ISBN: 1558994564 Notes: Includes bibliographical references and index. Subjects: Biomedical materials--Congresses. Poplymeric drug delivery systems--Congresses. Biomedical engineering--Congresses. Animal cell biotechnology--Congresses. Biomedical Engineering--methods--Congresses. Drug Delivery Systems--Congresses. Polymers--Congresses. Prostheses and Implants--Congresses. Tissue Transplantation--Congresses. Series: Materials Research Society symposia proceedings; v. 550 Variant Series: Materials Research Society symposium proceedings, 0272-9172; v. 550 LC Classification: R857.M3 B579 1999 Dewey Class No.: 610/.28 21

Biomedical modeling and simulation on a PC: a workbench for physiology and biomedical engineering / R.P. van Wijk van Brievingh, D.P.F. Möller, editors. Published/Created: New York: Springer-Verlag, c1993. Related Authors: Wijk van Brievingh, R. P. van. Möller, Dietmar. Description: xvi, 517 p.; ill.; 25 cm. + 6 computer disks (5 1/4 in.) ISBN: 0387976507 (acid-free) 3540976507 (Berlin: acid-free) Notes: System requirements for computer disks: IBM XT, AT, or PS/2 or compatible; 512K RAM; DOS 2.1 or later; hard disk with 3MB free space; video adapter and monitor capable of high resolution graphics (CGA, EGA, VGA, or Hercules); Epson FX-80 or IBM Pro Printer or compatible (recommended); mouse with Microsoft or Mouse Systems compatible driver (recommended). Includes bibliographical references and index. Subjects: Physiology--Computer simulation. Biotechnology--Computer simulation. Microcomputers. Series: Advances in simulation; v. 6 LC

Classification: QP33.6.D38 B68 1993 Dewey Class No.: 612/.001/13 20

Biomedical optical instrumentation and laser-assisted biotechnology / edited by A.M. Verga Scheggi ... [et al.]. Published/Created: Boston: Kluwer Academic, c1996. Related Authors: Verga Scheggi, A. M. (Anna Maria) Description: xxviii, 407 p.: ill.; 25 cm. ISBN: 0792341724 (alk. paper) Notes: Includes bibliographical references and index. Subjects: Optical instruments. Biomedical engineering. Lasers in biology. Biotechnology--Instruments. Series: NATO ASI series. Series E, Applied sciences; no. 325 LC Classification: R857.O6 B57 1996 Dewey Class No.: 610/.28 20

Biomedical science and technology: recent developments in the pharmaceutical and medical sciences / edited by A. Atilla Hincal and H. Süheyla Ka¸s. Published/Created: New York: Plenum Press, c1998. Related Authors: Hincal, A. Atilla. Ka¸s, H. Süheyla. International Symposium on Biomedical Science and Technology (4th: 1997: Istanbul, Turkey) Description: xii, 241 p.: ill.; 26 cm. ISBN: 0306458373 Notes: "Proceedings of the Fourth International Symposium on Biomedical Science and Technology, held September 15-17, 1997, in Istanbul, Turkey"--T.p. verso. Includes bibliographical references and index. Subjects: Biomedical engineering-- Congresses. Medical technology-- Congresses. Pharmaceutical technology- -Congresses. Biotechnology-- Congresses. Technology, Pharmaceutical--congresses. Technology, Medical--congresses. LC Classification: R856.A2 B577 1998

Dewey Class No.: 610/.28 21

Biomedical technology and public policy / edited by Robert H. Blank and Miriam K. Mills; prepared under the auspices of the Policy Studies Organization. Published/Created: New York: Greenwood Press, 1989. Related Authors: Blank, Robert H. Mills, Miriam K. Policy Studies Organization. Description: xv, 235 p.; 25 cm. ISBN: 0313266298 (lib. bdg.: alk. paper) Notes: Includes bibliographies and index. Subjects: Human reproductive technology--Government policy--United States. Medical technology-- Government policy--United States. Medical policy--United States. Bioethics. Biotechnology. Public Policy--United States. Series: Contributions in medical studies, 0886-8220; no. 26 LC Classification: RG133.5 .B56 1989 Dewey Class No.: 174/.2 20

Biomedical technology: innovations, the social consequences of science and technology program / BSCS. Published/Created: Dubuque, Iowa: Kendall/Hunt Pub. Co., c1984. Related Authors: Biological Sciences Curriculum Study. Description: ii, 89 p.: ill.; 28 cm. ISBN: 0840332866 (pbk.) Notes: Includes bibliographies. Subjects: Medical technology--United States--Social aspects. Biotechnology-- United States--Social aspects. LC Classification: R855.5.U6 B56 1984 Dewey Class No.: 362.1/042 19

Biomedical, biotechnology, and pharmaceutical innovation: Australia's opportunities. Published/Created: Double Bay, N.S.W.: CL Creations, 2000. Related Authors: Harrison, Roger. Description: 143 p.: col. ill.; 32 cm.

ISBN: 0646389270 0646389270 (hbk.) Summary: Produced in response to the need for Australian companies and research organisations to promote their capabilities on an international scale. Contains case studies covering leading edge companies and organisations and provides a unique reference point for students, scientists or organisations. Notes: Includes index. Subjects: Biotechnology--Australia. Biotechnology industries--Australia. Pharmaceutical industry--Australia. Medicine--Research--Australia. LC Classification: TP248.195.A8 B54 2000 Dewey Class No.: 338.4/76606/0994 21

Biopharmaceutical drug design and development / edited by Susanna Wu-Pong and Yongyut Rojanasakul; foreword by Joseph R. Robinson. Published/Created: Totowa, N.J.: Humana Press, c1999. Related Authors: Wu-Pong, Susanna. Rojanasakul, Yongyut. Description: xii, 435 p.: ill.; 24 cm. ISBN: 089603691X (acid-free paper) Notes: Includes bibliographical references and index. Subjects: Pharmaceutical biotechnology. Gene therapy. Biotechnology. Molecular biology. Biotechnology--methods. Molecular Biology. Genetic Engineering. LC Classification: RS380 .M65 1999 Dewey Class No.: 615/.19 21

Biopharmaceutical process validation / edited by Gail Sofer, Dane W. Zabriskie. Published/Created: New York: Marcel Dekker, c2000. Related Authors: Sofer, Gail. Zabriskie, Dane W., 1950- Description: xiv, 382 p.: ill.; 24 cm. ISBN: 0824702492 (alk. paper) Notes: Includes bibliographical references and index. Subjects: Pharmaceutical biotechnology--Quality

control. Series: Biotechnology and bioprocessing series LC Classification: RS380 .B5526 2000 Dewey Class No.: 615/.19 21

Biopharmaceuticals in the 1990s / edited by Gene R. Sacco, Chester A. Bisbee, Robin S. [i.e. J.] Rodgers. Published/Created: Burlington, Mass.: Decision Resources, Inc., 1991. Related Authors: Sacco, Gene R. Bisbee, Chester Allen. Rodgers, Robin J. Description: xi, 121 p.: ill.; 28 cm. Subjects: Pharmaceutical biotechnology. Pharmaceutical biotechnology industry. Biopharmaceutics Biotechnology Technology, Pharmaceutical Series: DR reports LC Classification: RS380 .B555 1991 Dewey Class No.: 615/.19 20

Biopharmaceuticals in transition / Industrial Biotechnology Association, Paine Webber, Bio/Technology. Published/Created: Woodlands, Tex.: Portfolio Pub. Co.; Houston: Gulf Pub. Co., Book Division, c1990. Related Authors: Paine Webber Inc. Industrial Biotechnology Association (U.S.). Meeting (8th: 1989: Washington, D.C.) Paine Webber-Bio/Technology Conference on Biopharmaceuticals Futures (1989: New York, N.Y.) Related Titles: [Bio/technology. Description: xvi, 400 p.: ill.; 24 cm. ISBN: 0943255120: Notes: Taken from Bio/technology magazine, from presentations at the Eighth Annual Meeting of the Industrial Biotechnology Association held in Washington, Oct. 11-13, 1989, and from panel discussions at the Paine Webber-Bio/Technology Conference on Biopharmaceuticals Futures held in New York, Sept. 13-15, 1989. Includes bibliographical references and index. Subjects:

Pharmaceutical biotechnology industry--United States Congresses. Pharmaceutical biotechnology industry--Europe--Congresses. Series: Advances in applied biotechnology series; v. 8 LC Classification: HD9666.5 .B56 1990 Dewey Class No.: 338.4/761519 20

Biopharmaceuticals, an industrial perspective / edited by Gary Walsh and Brendan Murphy. Published/Created: Dordrecht; Boston, Mass: Kluwer Academic, c1999. Related Authors: Walsh, Gary. Murphy, Brendan. Description: x, 514 p.: ill.; 25 cm. ISBN: 0792357469 (alk. paper) Notes: Includes bibliographical references and index. Subjects: Pharmaceutical biotechnology. Biological products--Therapeutic use. LC Classification: RS380 .B553 1999 Dewey Class No.: 615/.19 21

Bioprocessing safety: worker and community safety and health considerations / Warren C. Hyer, Jr., editor. Published/Created: Philadelphia, PA: ASTM, c1990. Related Authors: Hyer, Warren C., 1941- American Society for Testing and Materials. International Symposium on Large-Scale Bioprocessing Safety: Worker and Community Safety and Health Considerations (1st: 1987: Washington, D.C.) Description: iii, 173 p.: ill.; 23 cm. ISBN: 0803112645 Notes: Papers presented at the First International Symposium on Large-Scale Bioprocessing Safety: Worker and Community Safety and Health Considerations, held Oct. 6-8, 1987 in Washington, D.C. Includes bibliographical references. Subjects: Biotechnology--Safety Measures--Congresses. Biotechnology--Health aspects--Congresses. Series: ASTM special technical publication; 1051. Variant Series: STP; 1051 LC Classification: TP248.14 .B54 1990 Dewey Class No.: 660/.6/0289 20

Bioscan. Published/Created: Phoenix, AZ: Oryx Press, c1987- Description: v.; 30 cm. Vol. 1- Current Frequency: Bimonthly, Oct. 1993- Former Frequency: Annual, with 5 cumulative supplements, 1987- Aug. 1993 ISSN: 0887-6207 Cancel/Invalid LCCN: sn 86013810 Notes: Published: Atlanta, GA: American Health Consultants, Each update supercedes entirely the previous one. SERBIB/SERLOC merged record Also available on floppy disks. Subset of BioScan also available with AgriBioScan. AgriBioScan 1079-9060 (DLC) 95645859 (OCoLC)31620839 Subjects: Biotechnology industries--Directories. Biomedical Engineering--Directories. Technology--Directories. LC Classification: HD9999.B44 B56 Dewey Class No.: 338.7/6208/025 19

Biosensors & bioelectronics. Published/Created: Barking, Essex, England: Elsevier Applied Science, 1989- Description: v.: ill.; 24-28 cm. Vol. 5, no. 1- Current Frequency: 12 issues a year, <1997- Former Frequency: Six issues a year, 1989- 10 issues a year, <1994- Continues: Biosensors 0265-928X (DLC)sc 86002083 (OCoLC)11882549 ISSN: 0956-5663 Incorrect ISSN: 0265-928X Cancel/Invalid LCCN: sn 90031052 CODEN: BBIOE4 Notes: Title from cover. Published: Kidlington, Oxford, UK: Elsevier Science Ltd.,; Lausanne: Elsevier Science S.A., Includes a section called: SUBIS bibliography on biosensors and bioelectronics. SERBIB/SERLOC merged record

Indexed by: Bibliography of agriculture 0006-1530 Vol. 7, No. 8, 1992 Indx'd selectively by: Biological abstracts 0006-3169 1991- Chemical abstracts 0009-2258 1990- Computer & control abstracts 0036-8113 1990- Electrical & electronics abstracts 0036-8105 1990- Index medicus 0019-3879 1990- Physics abstracts 0036-8091 1990- Additional Form Avail.: Available also by subscription via the World Wide Web. Subjects: Biosensors--Periodicals. Bioelectronics--Periodicals. Biochemistry--methods--periodicals. Biotechnology--periodicals. Biosensors--abstracts. Biosensors--periodicals. Electronics, Medical--abstracts. Electronics, Medical--periodicals. LC Classification: R857.B54 B548 Dewey Class No.: 572 21

Biosensors: applications in medicine, environmental protection, and process control / edited by R.D. Schmid and F. Scheller. Published/Created: Weinheim, Federal Republic of Germany; New York, NY, USA: VCH, c1989. Related Authors: Schmid, Rolf, 1942- Scheller, F. (Frieder) GBF International Workshop on Biosensors (2nd: 1989: Braunschweig, Germany) Gesellschaft für Biotechnologische Forschung (Braunschweig, Germany) Description: xvii, 428 p.: ill.; 24 cm. ISBN: 0895739550 (U.S. alk. paper) Notes: Consists of papers presented at the GBF Workshop on Biosensors held in Braunschweig, Germany on May 22 and 23, 1989. Includes bibliographical references. Subjects: Biosensors--Congresses. Medical instruments and apparatus--Congresses. Environmental protection--Congresses. Biotechnological process control--Congresses. Biotechnological process monitoring--Congresses. Biosensors--congresses. Biotechnology--congresses. Series: GBF monographs, 0930-4320; v. 13 LC Classification: R857.B54 B553 1989 Dewey Class No.: 681/.761 20

Biotech '89: proceedings of the conference held in London, May 1989. Published/Created: London: Blenheim Online Publications, c1989. Related Authors: Blenheim Online Ltd. Related Titles: Biotech eighty nine. Description: x, 354 p.: ill.; 30 cm. ISBN: 0863531695 Notes: Includes bibliographical references. Subjects: Biotechnology industries--Congresses. Pharmaceutical industry--Congresses. LC Classification: HD9999.B442 B564 1989 Dewey Class No.: 338.4/76606 20

Biotech on fingertips. Published/Created: Terre Haute, IN: VPM & Associates, c1994- Description: v.; 28 cm. Vol. 1, no. 1 (Jan. 1994)- Current Frequency: Semiannual ISSN: 1075-2005 Cancel/Invalid LCCN: sn 94002132 Notes: SERBIB/SERLOC merged record Subjects: Pharmaceutical industry--Finance--Periodicals. Biotechnology industries--Finance--Periodicals. Pharmaceutical biotechnology--Periodicals. New products--Periodicals. Stock quotations--Periodicals. LC Classification: HD9665.1 .B56 Dewey Class No.: 660 12

Biotechnology advances. Published/Created: Oxford; New York: Pergamon Press, c1983- Description: v.: ill.; 23 cm. Some issues combined. Vol. 1, no. 1- Current Frequency: 8 times a year, 1999- Former Frequency: Two no. a year, 1983-1987 4 no. a year, 1988-1997 Bimonthly, 1998 ISSN: 0734-9750 Cancel/Invalid LCCN: sn 82007008 CODEN: BIADDD Notes:

Title from cover. Published: Tarrytown, NY: Elsevier Science Inc., 1996- SERBIB/SERLOC merged record Indexed by: Bibliography of agriculture 0006-1530 v. 15, no. 1, 1997. Indx'd selectively by: Life sciences collection 1985- Additional Form Avail.: Issued also in microform. Beginning with 1995, also available to subscribers, in PDF, via the World Wide Web. Subjects: Biotechnology--Periodicals. Biotechnology--Abstracts--Periodicals. Biochemistry--methods--periodicals. Biomedical Engineering--patents. Biomedical Engineering--periodicals. Patents--abstracts. Technology--periodicals. LC Classification: TP248.2 .B55 Dewey Class No.: 660/.6/05 19

Biotechnology and biopharmaceutical manufacturing, processing, and preservation / Kenneth E. Avis, Vincent L. Wu, editors. Published/Created: Buffalo Grove, IL: Interpharm Press, c1996. Related Authors: Avis, Kenneth E., 1918- Wu, Vincent L. Description: xiv, 386 p.: ill.; 24 cm. ISBN: 1574910167 (hbk.) Notes: Includes bibliographical references and indexes. Subjects: Biochemical engineering. Protein drugs--Storage. Protein drugs--Preservation. Pharmaceutical biotechnology. Series: Drug manufacturing technology series; v. 2 LC Classification: TP248.3 .B594 1996 Dewey Class No.: 615/.19 20

Biotechnology in clinical medicine / editors, Alberto Albertini, Claude Lenfant, Rodolfo Paoletti. Published/Created: New York: Raven Press, c1987. Related Authors: Albertini, Alberto. Lenfant, Claude. Paoletti, Rodolfo. Fondazione Giovanni Lorenzini. International Symposium on Biotechnology in Clinical Medicine (1987: Rome, Italy)

Description: xxi, 362 p.: ill.; 25 cm. ISBN: 0881673757 Notes: Papers presented at the International Symposium on Biotechnology in Clinical Medicine in Rome, Apr. 13-15, 1987, co-sponsored by the Fondazione Giovanni Lorenzini together with nine other scientific organizations. Includes bibliographies and index. Subjects: Biotechnology--Congresses. Medical innovations--Congresses. Genetic engineering--Congresses. Cardiovascular system--Diseases--Congresses. Vaccines--Congresses. Molecular biology--Congresses. Biotechnology--congresses. Medicine--congresses. LC Classification: TP248.14 .B567 1987 Dewey Class No.: 610/.28 19

Biotechnology of industrial antibiotics / edited by Erick J. Vandamme. Published/Created: New York: M. Dekker, c1984. Related Authors: Vandamme, Erick J., 1943- Description: xvii, 808 p.: ill.; 27 cm. ISBN: 0824770560 Notes: Includes bibliographies and index. Subjects: Antibiotics--Industrial applications. Biotechnology. Antibiotics. Technology, Pharmaceutical. Series: Drugs and the pharmaceutical sciences; v. 22 LC Classification: TP248.A62 B56 1984 Dewey Class No.: 660/.63 19

Biotechnology of vitamins, pigments, and growth factors / edited by Erick J. Vandamme. Published/Created: London; New York: Elsevier Applied Science, c1989. Related Authors: Vandamme, Erick J., 1943- Description: xii, 439 p.: ill.; 25 cm. ISBN: 1851663258 Notes: Includes bibliographies and index. Subjects: Vitamins--Biotechnology. Growth factors--Biotechnology. Pigments

(Biology)--Biotechnology.
Biotechnology. Growth Substances.
Pigments. Vitamins. Series: Elsevier
applied biotechnology series LC
Classification: TP248.65.V57 B56 1989
Dewey Class No.: 660/.63 20

China medical market report.
Published/Created: La Jolla, Calif.:
Golden Triangle Organization, Related
Authors: Golden Triangle Organization.
Description: v.; 28 cm. Notes:
Description based on: 1995; title from
cover. SERBIB/SERLOC merged
record Subjects: Pharmaceutical
industry--China--Periodicals. Medical
instruments and apparatus industry--
China Periodicals. Biotechnology
industries--China--Periodicals. LC
Classification: HD9672.C5 C47

Cray, William C. Miles, 1884-1984: a
centennial history / William C. Cray.
Published/Created: Englewood Cliffs,
N.J.: Prentice-Hall, c1984. Related
Titles: Miles. Description: viii, 277 p.:
ill. (some col.); 29 cm. ISBN:
0135830141 Notes: Spine Miles.
Includes index. Subjects: Miles
Laboratories--History. Pharmaceutical
industry--United States--History.
Biotechnology industries--United
States--History. LC Classification:
HD9999.B444 M553 1984 Dewey Class
No.: 338.7/616151/0973 19

Development of biopharmaceutical
parenteral dosage forms / edited by John
A. Bontempo. Published/Created: New
York: Marcel Dekker, 1997. Related
Authors: Bontempo, John A., 1930-
Description: ix, 248 p.: ill.; 24 cm.
ISBN: 082479981X (hardcover: alk.
paper) Notes: Includes bibliographical
references and index. Subjects:
Pharmaceutical biotechnology.

Parenteral solutions. Series: Drugs and
the pharmaceutical sciences; v. 85 LC
Classification: RS380 .D48 1997
Dewey Class No.: 615/.6 21

Drioli, E. Biocatalytic membrane reactors:
applications in biotechnology and the
pharmaceutical industry / Enrico Drioli,
Lidietta Giorno. Published/Created:
London; Philadelphia: Taylor &
Francis, c1999. Related Authors:
Giorno, Lidietta. Description: xx, 211
p.: ill.; 26 cm. ISBN: 0748406549
Notes: Includes bibliographical
references and index. Subjects:
Membrane reactors. Membranes,
Artificial. Bioreactors. Biotechnology.
Technology, Pharmaceutical. Membrane
reactors. Membrane separation. LC
Classification: TP248.25.M45 D75
1999 Dewey Class No.: 660/.2832 21

Enzyme technologies for pharmaceutical
and biotechnological applications /
edited by Herbert A. Kirst, Wu-Kuang
Yeh, Milton J. Zmijewski, Jr.
Published/Created: New York: Marcel
Dekker, c2001. Related Authors: Kirst,
Herbert A. Yeh, Wu-Kuang, 1942-
Zmijewski, Milton J., 1948-
Description: xv, 611 p.: ill.; 24 cm.
ISBN: 0824705491 (alk. paper) Notes:
Includes bibliographical references and
index. Subjects: Enzymes--
Biotechnology. Enzymes--Industrial
applications. Pharmaceutical
biotechnology. LC Classification:
TP248.65.E59 E5914 2001 Dewey
Class No.: 660.6/34 21

FCCSET Committee on Life Sciences and
Health. Biotechnology for the 21st
century: realizing the promise: a report /
by the Committee on Life Sciences and
Health. Published/Created:
[Washington, D.C.]: Federal

Coordinating Council for Science, Engineering, and Technology, [1993] Related Titles: Biotechnology for the twenty-first century. Biotec. Description: ix, 90 p.: ill. (some col.); 28 cm. Notes: "A supplement to the President's fiscal year 1994 budget." Shipping list no.: 93-0515-P. "Biotec"--Cover. "June 30, 1993"--Cover. Includes bibliographical references. Subjects: Biotechnology--Research--United States. LC Classification: TP248.185 .F33 1993 Dewey Class No.: 660/.6/072073 20

FCCSET Committee on Life Sciences and Health. Biotechnology for the 21st century: a report / by the FCCSET Committee on Life Sciences and Health. Published/Created: [Washington, DC]: The Committee: G.P.O., [1992] Description: 125 p.: ill. (some col.); 28 cm. ISBN: 016036101X Notes: "February 1992." "FY 1993." Subjects: Biotechnology--Research--United States. LC Classification: TP248.185 .F33 1992 Dewey Class No.: 660/.6/072073 20

Filtration in the biopharmaceutical industry / edited by Theodore H. Meltzer, Maik W. Jornitz. Published/Created: New York: Marcel Dekker, c1998. Related Authors: Meltzer, Theodore H. Jornitz, Maik W., 1961- Description: xiv, 933 p.: ill.; 27 cm. ISBN: 0824798961 Notes: Includes bibliographical references and index. Subjects: Biological products--Separation. Filters and filtration. Pharmaceutical biotechnology--Methodology. LC Classification: RS190.B55 F54 1998 Dewey Class No.: 660/.284245 21

Frontiers in biomedicine and biotechnology. Published/Created: Mount Prospect, IL,

U.S.A.: ATL Press, c1993- Related Authors: Symposium on Industrial Polysaccharides. Description: v.: ill. Vol. 1- Current Frequency: Semiannual ISSN: 1067-1897 Cancel/Invalid LCCN: sn 92006910 CODEN: FBBIET Notes: Issues for Vol. 1 (1993)- contain proceedings of the Symposium on Industrial Polysaccharides. Indx'd selectively by: Chemical abstracts 0009-2258 Subjects: Biochemistry--methods--periodicals.

Genetically engineered human therapeutic drugs / [compiled by] David N. Copsey, Sabine Y.J. Delnatte. Published/Created: New York, N.Y.: Stockton Press, c1988. Related Authors: Copsey, David N. Delnatte, Sabine Y. J. Description: viii, 675 p.; 30 cm. ISBN: 0935859284: Notes: Includes bibliographies and index. Subjects: Recombinant molecules--Therapeutic use--Handbooks, manuals, etc. Pharmaceutical biotechnology--Handbooks, manuals, etc. Genetic engineering--Handbooks, manuals, etc. LC Classification: RM666.R37 G46 1988 Dewey Class No.: 615/.19 19

Haider, Syed Imtiaz. Validation standard operating procedures: a step-by-step guide for achieving compliance in the pharmaceutical, medical device, and biotech industry / by Syed Imitiaz Haider. Published/Created: Boca Raton, FL: St. Lucie Press, 2001. Projected Pub. Date: 0112 Description: p. cm. ISBN: 1574443313 Notes: Includes bibliographical references. Subjects: Pharmaceutical technology--Quality control. Pharmaceutical industry--Standards--United States. Biotechnology industries--Standards--United States. Medical instruments and apparatus industry--Standards United

States. LC Classification: RS192 .H353 2001 Dewey Class No.: 681/.761/021873 21

Houghton, Peter J. Laboratory handbook for the fractionation of natural extracts / Peter J. Houghton and Amala Raman. Edition Information: 1st ed. Published/Created: London; New York: Chapman & Hall, 1998. Related Authors: Raman, Amala. Description: vi, 199 p.: ill.; 24 cm. ISBN: 0412749106 Notes: Includes bibliographical references and index. Subjects: Plant bioactive compounds-- Separation. Plant biotechnology. Medicinal plants. Plant Extracts-- analysis. Fractionation--methods. LC Classification: QK898.B54 H68 1998 Dewey Class No.: 660.6 21

Human fertility, health, and food: impact of molecular biology and biotechnology / edited by David Puett. Published/Created: New York: United Nations Fund for Population Activities, 1984. Related Authors: Puett, David. United Nations Fund for Population Activities. Description: xvii, 254 p.: ill.; 26 cm. Notes: Includes bibliographies. Subjects: Biotechnology. Molecular biology. Health. Fertility, Human. Nutrition. LC Classification: TP248.2 .H86 1984 Dewey Class No.: 616.6/92 20

Huxsoll, Jean F. Quality assurance for biopharmaceuticals / Jean F. Huxsoll. Published/Created: New York: Wiley, c1994. Description: 206 p.: ill.; 24 cm. ISBN: 0471036560 (alk. paper) Notes: "A Wiley-Interscience publication". Includes bibliographical references and index. Subjects: Pharmaceutical biotechnology--Quality control. LC Classification: RS380 .H88 1994

Dewey Class No.: 615/.19 20

Impact of biotechnology on specialty chemicals & pharmaceutical markets. Published/Created: Norwalk, Conn., U.S.A. (6 Prowitt St., Norwalk 06855): International Resource Development, c1985. Related Authors: International Resource Development, inc. Description: v, 208 leaves; 28 cm. Notes: "July 1985." Subjects: Pharmaceutical industry--Technological innovations. Chemical industry-- Technological innovations. Biotechnology. Market surveys. Series: Report (International Resource Development, inc.); #657. Variant Series: Report / International Resource Development, Inc.; #657 LC Classification: HD9665.5 .I48 1985 Dewey Class No.: 338.4/566 19

International Symposium of Tissue Engineering for Therapeutic Use (1st: 1997: Kyoto, Japan) Tissue engineering for therapeutic use 1: proceedings of the First International Symposium of Tissue Engineering for Therapeutic Use, Kyoto, Japan, 2 March 1997 / editor, Yoshito Ikada; co-editor, Yoshio Yamaoka. Edition Information: 1st ed. Published/Created: Amsterdam; New York: Elsevier, 1998. Related Authors: Ikada, Yoshito, 1935- Yamaoka, Yoshio. Description: xxviii, 159 p.: ill. (some col.); 25 cm. ISBN: 0444829938 (alk. paper) Notes: Includes bibliographical references and indexes. Subjects: Animal cell biotechnology-- Congresses. Regeneration (Biology)-- Congresses. Tissue Culture--congresses. Regeneration--congresses. Biomedical Engineering--congresses. Series: International congress series; no. 1169. Variant Series: Excerpta Medica international congress series; 1169 LC

Classification: TP248.27.A53 I586 1997 Dewey Class No.: 610/.28 21

International Symposium of Tissue Engineering for Therapeutic Use (2nd: 1997: Tokyo, Japan) Tissue engineering for therapeutic use 2: proceedings of the Second International Symposium of Tissue Engineering for Therapeutic Use, Tokyo, 30-31 October 1997 / editor, Yoshito Ikada; co-editor, Shoji Enomoto. Edition Information: 1st ed. Published/Created: Amsterdam; New York: Elsevier, 1998. Related Authors: Ikada, Yoshito, 1935- Enomoto, Sh¯oji, 1936- Description: viii, 137 p.: ill.; 25 cm. ISBN: 0444500766 (alk. paper) Notes: Includes bibliographical references and index. Subjects: Animal cell biotechnology--Congresses. Regeneration (Biology)--Congresses. Tissue culture--Congresses. Series: International congress series; no. 1170. Variant Series: Excerpta Medica international congress series; 1170 LC Classification: TP248.27.A53 I586 1997a Dewey Class No.: 660.6 21

International Symposium of Tissue Engineering for Therapeutic Use (3rd: 1998: Tokyo, Japan) Tissue engineering for therapeutic use 3: proceedings of the Third International Symposium of Tissue Engineering for Therapeutic Use, Tokyo, 4-5 September 1998 / editor, Yoshito Ikada; co-editor, Teruo Okano. Edition Information: 1st ed. Published/Created: Amsterdam; New York: Elsevier, 1999. Related Authors: Ikada, Yoshito, 1935- Okano, Teruo. Description: viii, 161 p.: ill. (some col.); 25 cm. ISBN: 0444500294 (acid-free paper) Notes: Includes bibliographical references and indexes. Subjects: Animal cell biotechnology--Congresses. Regeneration (Biology)--Congresses.

Biomedical engineering--Congresses. Tissue Culture--congresses. Regeneration--congresses. Biomedical Engineering--congresses. Series: International congress series; no. 1175 LC Classification: TP248.27.A53 I586 1998 Dewey Class No.: 660.6 21

International Symposium of Tissue Engineering for Therapeutic Use (4th: 1999: Kyoto, Japan). Tissue engineering for therapeutic use 4: proceedings of the Fourth International Symposium of Tissue Engineering for Therapeutic Use, Tokyo, 23rd-24th September 1999 / editor, Yoshito Ikada; co-editor, Yoshihiko Shimizu. Edition Information: 1st ed. Published/Created: Amsterdam; New York: Elsevier, 2000. Related Authors: Ikada, Yoshito, 1935- Shimizu, Yoshihiko. Description: viii, 192 p.: ill. (some col.), charts; 25 cm. ISBN: 0444502939 Notes: Includes bibliographical references and indexes. Subjects: Animal cell biotechnology--Congresses. Regeneration (Biology)--Congresses. Biomedical engineering--Congresses. Regeneration (Biology)--Congresses. Biomedical engineering--Congresses. Animal cell biotechnology--Congresses. Tissue Culture--Congresses. Biomedical Engineering. Regeneration--Congresses. Series: International congress series; no. 1198 LC Classification: TP248.27.A53 I586 1999 Dewey Class No.: 660.6 21

International Symposium of Tissue Engineering for Therapeutic Use (5th: 2000: Tsukuba, Japan) Tissue engineering for therapeutics use 5: proceedings of the Fifth International Symposium on Tissue Engineering for Therapeutic Use held in Tsukuba, Japan, 16-17 November 2000 / editor, Yoshito Ikada; co-editor, Norio

Ohshima. Published/Created: Amsterdam; New York: Elsevier Science Ltd., 2001. Related Authors: Ikada, Yoshito, 1935- Ohshima, Norio. Description: viii, 226 p.: ill. (some col.); 25 cm. ISBN: 0444505547 (hardcover: alk. paper) Notes: Includes bibliographical references and index. Subjects: Animal cell biotechnology--Congresses. Biomedical engineering--Congresses. Series: International congress series, 0531-5131; 1222 LC Classification: TP248.27.A53 I586 2000 Dewey Class No.: 660.6 21

Isolator technology: applications in the pharmaceutical and biotechnology industries / edited by Carmen M. Wagner, James E. Akers. Published/Created: Buffalo Grove, IL: Interpharm Press, c1995. Related Authors: Wagner, Carmen M. Akers, James E. Description: xxi, 384 p.: ill.; 24 cm. ISBN: 0935184783 Notes: Includes bibliographical references and indexes. Subjects: Pharmaceutical biotechnology. Contamination (Technology) Clean rooms. Drugs--Sterilization. LC Classification: RS380 .I86 1995 Dewey Class No.: 615/.19 20

Kanarek, Alex D. The bioprocessing industry: growing markets in BioPharm manufacturing technology / Alex D. Kanarek; managing editor, Susan C. DiClemente. Published/Created: Westborough, MA: D&MD Reports, c2000. Related Authors: DiClemente, Susan C. Description: vii, [446] p.: ill.; 29 cm. ISBN: 1579361544 Subjects: Pharmaceutical biotechnology industry. Pharmaceutical biotechnology industry--Directories. Biotechnology industries. Biotechnology industries--Directories. Market surveys. Series: D & MD reports; #9010. Variant Series: D&MD Reports: #9010 LC Classification: HD9665.5 .K36 2000

Krul, Kenneth G. Outlook for biopharmaceutical technologies to 2005 / Kenneth G. Krul. Published/Created: Waltham, MA: Decision Resources, [c1996] Related Authors: Decision Resources, Inc. Description: x, 108 leaves: ill.; 30 cm. Subjects: Biotechnology industries--United States. Biopharmaceutics--United States. Biotechnology--United States. Genetic engineering industry--United States. Technology, Pharmaceutical--United States. Market surveys--United States. Series: DR reports LC Classification: HD9999.B443 U648 1996

Krul, Kenneth G. Outlook for biopharmaceutical technologies to 2005 / Kenneth G. Krul. Published/Created: Waltham, MA: Decision Resources, [c1996] Related Authors: Decision Resources, Inc. Description: x, 108 leaves: ill.; 30 cm. Subjects: Biotechnology industries--United States. Biopharmaceutics--United States. Biotechnology--United States. Genetic engineering industry--United States. Technology, Pharmaceutical--United States. Market surveys--United States. Series: DR reports LC Classification: HD9999.B443 U648 1996

Kuhl, Phillips. Commercialization of antibody engineering / Phillips L. Kuhl. Published/Created: Burlington, Mass.: Decision Resources, c1991. Description: xv, 281 p.: ill.; 28 cm. Notes: "December 1991." Subjects: Immunotherapy equipment industry. Biotechnology industries. Immunotechnology. Biomedical engineering. Market surveys. Series: DR reports LC Classification: HD9995.I452 K84 1991

Kuhl, Phillips. Commercialization of antibody engineering / Phillips L. Kuhl. Published/Created: Burlington, Mass.: Decision Resources, c1991. Description: xv, 281 p.: ill.; 28 cm. Notes: "December 1991." Subjects: Immunotherapy equipment industry. Biotechnology industries. Immunotechnology. Biomedical engineering. Market surveys. Series: DR reports LC Classification: HD9995.I452 K84 1991

Lee, Chi-Jen. Managing biotechnology in drug development / Chi-Jen Lee. Published/Created: Boca Raton: CRC Press, c1996. Description: 181 p.: ill.; 25 cm. ISBN: 084939466X (acid-free paper) Notes: Includes bibliographical references and index. Subjects: Pharmaceutical biotechnology. LC

Martineau, William D. Biotechnology in health care / project director, William Martineau, T. Kevin Swift. Published/Created: Cleveland, Ohio: Freedonia Group, c1988. Related Authors: Swift, T. Kevin. Description: v, 99 leaves; 28 cm. Notes: "June 1988." Includes bibliographical references (leaves 98-99). Subjects: Pharmaceutical industry--United States. Diagnostic reagents industry--United States. Biotechnology industries-- United States. Biotechnology-- Research--United States. Market surveys--United States. Series: Business research report (Cleveland, Ohio); B52. Variant Series: Business research report; B52 LC Classification: HD9666.5 .M38 1988

Maulik, Sunil, 1960- Molecular biotechnology: therapeutic applications and strategies / Sunil Maulik, Salil D. Patel. Published/Created: New York: Wiley-Liss, c1997. Related Authors: Patel, Salil D. Description: xx, 223 p.: ill.; 24 cm. ISBN: 0471116815 (pbk.: alk. paper) Notes: Includes bibliographical references and index. Subjects: Genetic engineering. Molecular biology. Biotechnology. LC Classification: TP248.6 .M36 1997 Dewey Class No.: 615 20

Microbes: for health, wealth & sustainable environment / editor, Ajit Varma. Published/Created: New Delhi: Malhotra Pub. House, 1998. Related Authors: Varma, A. (Ajit), 1939- Description: xv, 826, xxx p.: ill.; 24 cm. ISBN: 818504838X Notes: Includes bibliographical references and index. Subjects: Agricultural microbiology. Industrial microbiology. Medical microbiology. Microbial biotechnology. LC Classification: QR51 .M43 1998 Dewey Class No.: 660.6/2 21

Odum, Jeffery N. Sterile product facility design and project management / Jeffery N. Odum. Published/Created: Buffalo Grove, Ill.: Interpharm Press, c1997. Description: xv, 450 p.: ill.; 30 cm. ISBN: 1574910205 Notes: Includes bibliographical references and index. Subjects: Biotechnology laboratories-- Design and construction. Pharmaceutical industry. Clean rooms-- Design and construction. LC Classification: TP248.24 .O38 1997 Dewey Class No.: 660/.6/078 20

Pharmaceutical & biotech daily. Published/Created: Washington, DC: King Pub. Group, c1994. Description: 1 v.; 28 cm. Vol. 1, no. 1 (Jan. 25, 1994)-

v. 1, issue 187 (Nov. 7 1994). Current Frequency: Daily (Monday through Friday) Absorbed by: Washington drug letter 0194-1291 (DLC) 81643726 (OCoLC)5320071 ISSN: 1074-8636 Incorrect ISSN: 1067-1196 Cancel/Invalid LCCN: sn 94001341 CODEN: PBDAEN Notes: Title from caption. SERBIB/SERLOC merged record Merger of: Biotech daily, and: Pharmaceutical daily. Subjects: Biotechnology--Periodicals. Biotechnology industries--Periodicals. Pharmaceutical industry--Periodicals. Pharmaceutical biotechnology--Periodicals. LC Classification: TP248.13 .P47 Dewey Class No.: 338.4/766/06 20

Pharmaceutical biotechnology: a programmed text / edited by S. William Zito. Edition Information: 2nd ed. Published/Created: Lancaster, Pa.: Technomic Pub. Co., c1997. Related Authors: Zito, S. William. Description: xii, 282 p.: ill.; 26 cm. ISBN: 1566765196 Notes: Includes bibliographical references and index. Subjects: Pharmaceutical biotechnology. LC Classification: RS380 .P48 1997 Dewey Class No.: 615/.19

Pharmaceutical biotechnology: a programmed text / edited by S. William Zito. Published/Created: Lancaster, Pa.: Technomic Pub. Co., c1992. Related Authors: Zito, S. William. Description: x, 177 p.: ill.; 25 cm. ISBN: 0877629110 Notes: Includes bibliographical references and index. Subjects: Pharmaceutical biotechnology. LC Classification: RS380 .P48 1992 Dewey Class No.: 615/.19 20

Pharmaceutical biotechnology: an introduction for pharmacists and pharmaceutical scientists / edited by Daan J.A. Crommelin and Robert D. Sindelar. Published/Created: Amsterdam, The Netherlands: Harwood Academic Publishers, c1997. Related Authors: Crommelin, D. J. A. (Daan J. A.) Sindelar, Robert D. Description: xxviii, 369 p.: ill. (some col.); 29 cm. ISBN: 9057022486 (hardcover) Notes: Includes bibliographical references and index. Subjects: Pharmaceutical biotechnology. Technology, Pharmaceutical. LC Classification: RS380 .P484 1997 Dewey Class No.: 615/.19 21

Pharmaceutical industry guide. Published/Created: Research Triangle Park, NC: Institute for Biotechnology Information, c1995-c1997. Related Authors: Institute for Biotechnology Information (North Carolina Biotechnology Center) Description: 2 v.; 28 cm. [1st ed.] 1995-96-2nd ed., 1997. Current Frequency: Annual ISSN: 1085-4185 Notes: "Drug companies, biotech firms & CROs." SERBIB/SERLOC merged record Subjects: Pharmaceutical industry--United States--Directories. Biotechnology industries--United States--Directories. Research, Industrial--United States--Directories. Biotechnology--United States--directories. Drug Industry--United States--directories. Research--United States--directories. LC Classification: HD9666.3 .P13 Dewey Class No.: 338.7/616151/02573 20

Pharmaceutical innovation: revolutionizing human health / edited by Ralph Landau, Basil Achilladelis, and Alexander Scriabine. Published/Created:

Philadelphia: Chemical Heritage Press, c1999. Related Authors: Landau, Ralph. Achilladelis, Basil, 1937- Scriabine, Alexander. Description: xxiii, 408 p.: ill.; 24 cm. ISBN: 0941901211 Notes: Includes bibliographical references and index. Subjects: Pharmaceutical industry--Technological innovations. Pharmaceutical biotechnology. Drug development. LC Classification: RS122 .P42 1999 Dewey Class No.: 615/.19 21

Pharmaceutical medicine, biotechnology, and European law / edited by Richard Goldberg and Julian Lonbay. Published/Created: Cambridge [England]; New York: Cambridge University Press, 2000. Related Authors: Goldberg, Richard. Lonbay, Julian. Description: xxxv, 241 p.; 23 cm. ISBN: 0521792495 (hb) Notes: Includes index. Subjects: Pharmacy--Law and legislation--Europe. Drugs--Law and legislation--Europe. Biotechnology industries--Law and legislation--Europe. Medical laws and legislation--Europe. LC Classification: KJC6191 .P48 2000 Dewey Class No.: 344/.04233/094 21

Pharmaceuticals/biotechnology market sourcebook / Frost & Sullivan. Published/Created: Mountain View, Calif.: Frost & Sullivan, Related Authors: Frost & Sullivan. Description: v.: ill.; 28 cm. Current Frequency: Annual Notes: Description based on: 1994 ed. Subjects: Pharmaceutical industry--Periodicals. Biotechnology industries--Periodicals. Pharmaceutical biotechnology industry--Periodicals. Veterinary supplies industry--Periodicals. Market surveys--Periodicals. LC Classification: HD9665.1 .P455

Polysaccharides in medicinal applications / edited by Severian Dumitriu. Published/Created: New York: M. Dekker, c1996. Related Authors: Dumitriu, Severian, 1939- Description: viii, 794 p.: ill.; 26 cm. ISBN: 0824795407 (hardcover: alk. paper) Notes: Includes bibliographical references and index. Subjects: Polysaccharides--Physiological effect. Polysaccharides--Therapeutic use. Polysaccharides--Biotechnology. Polysaccharides--chemical synthesis. Polysaccharides--therapeutic use. Glycoconjugates. Glycosides. Enzyme Stability. Biotechnology. LC Classification: QP702.P6 P645 1996 Dewey Class No.: 610/.28 20

Scott, Trevor. The dolomite solution / Trevor Scott. Edition Information: 1st ed. Published/Created: Bend, Or.: Salvo Press, c2000. Description: 202 p.; 23 cm. ISBN: 0966452070 (paperback: alk. paper) Subjects: Private investigators--Fiction. Pharmaceutical biotechnology industry--Fiction. Genetic engineering--Fiction. Genre/Form: Adventure fiction. LC Classification: PS3569.C6787 D65 2000 Dewey Class No.: 813/.54 21

Struck, Mark-Michael. Global biotechnology product registration: E.U., U.S., and Japan / Mark-Michael Struck, Hiromi Okabe, Mark Mathieu. Published/Created: Waltham, MA: Parexel International Corp., 1997. Related Authors: Okabe, Hiromi. Mathieu, Mark P. Description: xvii, 392 p.: ill.; 28 cm. ISBN: 1882615328 Notes: Includes bibliographical references. Subjects: Pharmaceutical biotechnology. Biotechnology industries--Marketing. Pharmaceutical biotechnology--Law and legislation--Europe. Pharmaceutical biotechnology--

Law and legislation--United States. Pharmaceutical biotechnology--Law and legislation--Japan. LC Classification: RS380 .S77 1997 Dewey Class No.: 615/.19 21

Synthetic polymers for biotechnology and medicine / [edited by] Ruth Freitag. Published/Created: Georgetown, TX: Landes Bioscience; Austin, TX: Eurekah.com, 2001. Projected Pub. Date: 0101 Related Authors: Freitag, Ruth, 1961- Description: p.; cm. ISBN: 1587060272 (alk. paper) Notes: Includes bibliographical references and index. Subjects: Polymers in medicine. Polymers. Biomedical Engineering. Biotechnology. Equipment Design. Series: Biotechnology intelligence unit (Unnumbered) Variant Series: Biotechnology intelligence unit LC Classification: R857.P6 S975 2001 Dewey Class No.: 610/.28 21

The BioTech pages. Published/Created: Lake Forest, Calif.: Biomedical Co. Intelligence, [c1993- Related Authors: Biomedical Company Intelligence. Description: v.; 27 cm. c1993- Current Frequency: Annual with semiannual update ISSN: 1068-6819 Cancel/Invalid LCCN: sn 93005963 Notes: SERBIB/SERLOC merged record Subjects: Biotechnology industries-- United States--Directories. Medical instruments and apparatus industry-- United States Directories. Pharmaceutical industry--United States--Directories. LC Classification: HD9999.B443 U6145

The Impact of biotechnology on autoimmunity / edited by A.G. Dalgleish, A. Albertini, and R. Paoletti. Published/Created: Dordrecht; Boston: Kluwer Academic Publishers; Milan,

Italy: Fondazione Giovanni Lorenzini; Houston, Tex., U.S.A.: Giovanni Lorenzini Medical Foundation, c1994. Related Authors: Dalgleish, A. G. (Angus G.) Albertini, Alberto. Paoletti, Rodolfo. Description: x, 136 p.: ill.; 25 cm. ISBN: 0792327241 (hbk.: acid-free paper) Notes: "This volume contains the chapters that summarize the plenary presentations given at the ... meeting in Florence, Italy in June 1993"--Pref. Includes bibliographical references and index. Subjects: Autoimmune diseases-- Congresses. Autoimmunity-- Congresses. Biotechnology-- Congresses. Immunotechnology-- Congresses. Autoimmune Diseases-- diagnosis. Autoimmune Diseases-- therapy. Autoimmunity--immunology. Series: Medical science symposia series; v. 6 LC Classification: RC600 .I49 1994 Dewey Class No.: 616.97/8 20

The Impact of hybridoma technology on the medical device and diagnostic product industry: proceedings of the educational seminar, Stouffer's National Center Hotel, Arlington, Virginia, June 14-15, 1982 / sponsored by the Health Industry Manufacturers Association in cooperation with the Food and Drug Administration, the Office of Technology Assessment, and Washington University, St. Louis; edited by Timothy J. Henry. Published/Created: Washington, D.C.: HIMA, c1982. Related Authors: Henry, Timothy J. Health Industry Manufacturers Association. Description: xxi, 260 p.: ill.; 28 cm. Subjects: Biotechnology--Congresses. Biotechnology industries--Congresses. Hybridomas--Congresses. Monoclonal antibodies--Congresses. Series: HIMA report; 82-1 LC Classification: TP248.14 .I47 1982 Dewey Class No.:

338.4/761028 19

The job bank guide to health care companies. Published/Created: Holbrook, Mass.: Adams Media Corp., c1998- Description: v.; 23 cm. First issue carries only copyright date. 1998- Current Frequency: Annual ISSN: 1099-0208 Cancel/Invalid LCCN: sn 98004382 Notes: SERBIB/SERLOC merged record Subjects: Health facilities--United States--Directories. Medical laboratories--United States--Directories. Pharmaceutical industry--United States--Directories. Biotechnology industry--United States--Directories. Series: Job bank series. LC Classification: RA977 .J63 Dewey Class No.: 362.1/025/73 21

The pharmaceutical companies fact file. Running Fact file Published/Created: Richmond, Surrey: PJB Publications Ltd, 1994- Related Authors: PJB Publications Ltd. Description: v.; 30 cm. 1994 issued in 3 v.; 1995- issued in 4 v. 1996-<1997 also called 3rd ed.-<4th ed. 1994- Current Frequency: Annual Continued by: Scrip pharmaceutical companies fact file (DLC) 2002242002 (OCoLC)48962213 Cancel/Invalid LCCN: sn 97039209 Notes: SERBIB/SERLOC merged record Additional Form Avail.: Also available as a quarterly-updated CD-ROM, <1996- Subjects: Pharmaceutical industry--Directories. Pharmaceutical industry--Finance--Directories. Biotechnology industries--Directories. Biotechnology industries--Finance--Directories. Drugs--Research--Directories. Drugs--Testing--Directories. Series: Scrip reports (Series) Variant Series: Scrip reports LC Classification: HD9665.3 .P48 Dewey

Class No.: 338.7/616151/025 21

The Status and impact of biotechnology in the U.S. Published/Created: San Jose, CA, USA (4340 Stevens Creek Blvd., San Jose 95129): Creative Strategies International, c1983. Related Authors: Creative Strategies International. Description: ii, 111 p.: ill.; 28 cm. Notes: "December 1983." Subjects: Biotechnology--United States. Series: Pharmaceuticals and biotechnology industry analysis service LC Classification: TP248.2 .S73 1983 Dewey Class No.: 574.87/3282 19

United States. Congress. Senate. Committee on Health, Education, Labor, and Pensions. The future of food: biotechnology and consumer confidence: hearing of the Committee on Health, Education, Labor, and Pensions, United States Senate, One Hundred Sixth Congress, second session, on examining issues related to biotechnology and genetically engineered food, and the measures needed to ensure consumer safety and confidence with respect to these food products, September 26, 2000. Published/Created: Washington: U.S. G.P.O.: For sale by the Supt. of Docs., Congressional Sales Office, U.S. G.P.O., 2000. Description: iii, 71 p.; 23 cm. ISBN: 0160645603 Notes: "Printed for the use of the Committee on Health, Education, Labor, and Pensions." Distributed to some depository libraries in microfiche. Shipping list no.: 2001-0088-P. Includes bibliographical references (p. 60-62). Subjects: Food--Biotechnology--United States. Genetically modified foods--United States. Food adulteration and inspection--United States. Consumer protection--United States. Series: United

States. Congress. Senate. S. hrg.; 106-698. Variant Series: S. hrg.; 106-698 LC Classification: KF26 .L27+

United States. Congress. Senate. Committee on Labor and Human Resources. The FDA and the future of American biomedical and food industries: hearing of the Committee on Labor and Human Resources, United States Senate, One Hundred Fourth Congress, first session, on examining activities of the Food and Drug Afministration focusing on the challenges and opportunities facing the pharmaceutical, biotech, medical device, and food industries, and FDA's regulation of these industries, April 5 and 6, 1995. Published/Created: Washington: U.S. G.P.O.: For sale by the U.S. G.P.O., Supt. of Docs., Congressional Sales Office, 1995. Description: iv, 202 p.: ill.; 24 cm. ISBN: 0160472784 Notes: Distributed to some depository libraries in microfiche. Shipping list no.: 95-0243-P. Includes bibliographical references. Subjects: United States. Food and Drug Administration. Food industry and trade--Technological innovations--United States. Pharmaceutical industry--Technological innovations--United States. Biotechnology industries--Technological innovations United States. Series: United States. Congress. Senate. S. hrg.; 104-45. Variant Series: S. hrg.; 104-45 LC Classification: KF26 .L27 1995

Werth, Barry. The billion-dollar molecule: one company's quest for the perfect drug / Barry Werth. Published/Created: New York: Simon & Schuster, c1994. Description: 445 p.; 25 cm. ISBN: 0671723278: Notes: Includes bibliographical references (p. [423]-432) and index. Subjects: Vertex

Pharmaceuticals Incorporated. Pharmaceutical biotechnology. Pharmaceutical industry--United States. LC Classification: RS380 .W47 1994 Dewey Class No.: 615/.19 20

Whitmore, Elaine. Product development planning for health care products regulated by the FDA / Elaine Whitmore. Published/Created: Milwaukee, Wis.: ASQC Quality Press, c1997. Description: xv, 159 p.: ill.; 24 cm. ISBN: 0873894162 (acid-free paper) Notes: Includes bibliographical references and index. Subjects: United States. Food and Drug Administration. Pharmaceutical industry--United States. Medical instruments and apparatus industry--United States. Biotechnology industries--United States. New products--United States. LC Classification: HD9666.5 .W48 1997 Dewey Class No.: 681/.761/0685 21

WHO Symposium on Plants and Health for All: Scientific Advancement (1st: 1991: K͞obe-shi, Japan) Natural resources and human health: plants of medicinal and nutritional value / editors, Shigeaki Baba, Olayiwola Akerele, Yuji Kawaguchi. Published/Created: Amsterdam; New York: Elsevier, 1992. Related Authors: Baba, Shigeaki, 1925- Akerele, Olayiwola. Kawaguchi, Yuji. Description: xviii, 227 p.: ill.; 25 cm. ISBN: 0444894071 (alk. paper) Notes: "Proceedings of the 1st WHO Symposium on Plants and Health for All: Scientific Advancement, Kobe, Japan, 26-28 August 1991." Includes bibliographical references and index. Subjects: Materia, medica, Vegetable--Congresses. Plant biotechnology--Congresses. Plants, Edible--Congresses. Conservation of Natural Resources--congresses. Health Promotion--

congresses. Plants, Edible--congresses.
Plants, Medicinal--congresses. LC
Classification: RS164 .W45 1991
Dewey Class No.: 615/.32 20

AGRICULTURE

Advanced engineered pesticides / edited by Leo Kim. Published/Created: New York: Marcel Dekker, c1993. Related Authors: Kim, Leo, 1942- Description: xii, 430 p.: ill.; 24 cm. ISBN: 0824789903 (alk. paper) Notes: Includes bibliographical references and index. Subjects: Microbial pesticides. Plants--Disease and pest resistance-- Genetic aspects. Crops--Genetic engineering. Agricultural biotechnology. LC Classification: SB976.M55 A38 1993 Dewey Class No.: 632/.96 20

Advances in plant biotechnology / edited by D.D.Y. Ryu, S. Furusaki. Published/Created: Amsterdam; New York: Elsevier, 1994. Related Authors: Ryu, D. D. Y. (Dewey D. Y.) Furusaki, S. (Shintar̄o), 1938- Description: xvi, 373 p.: ill.; 25 cm. ISBN: 0444899391 Notes: Includes bibliographical references and index. Subjects: Plant biotechnology. Plants biotechnology Series: Studies in plant science; 4 LC Classification: TP248.27.P55 A38 1994 Dewey Class No.: 660/.6 20

AgBiotech news and information. Published/Created: Wallingford, Oxon, UK: C.A.B. International, 1989- Related Authors: C.A.B. International. Related Titles: [CAB abstracts. Description: v.; 30 cm. Vol. 1, no. 1 (Feb. 1989)- Current Frequency: Monthly Former Frequency: Bimonthly ISSN: 0954- 9897 Notes: Title from cover. "A monthly journal with news, reviews and abstracts compiled from the CAB abstracts database." SERBIB/SERLOC merged record Subjects: Agricultural biotechnology--Periodicals. Agricultural biotechnology--Abstracts--Periodicals. Agriculture--methods--abstracts. Agriculture--methods--periodicals. Biotechnology--abstracts. Biotechnology--periodicals. LC Classification: S494.5.B563 A36 Dewey Class No.: 630 20

AgriBioScan: the agricultural biotechnology directory. Published/Created: Phoenix, Ariz.: Oryx Press, c1995- Description: v.; 28 cm. Vol. 1 (Apr. 1995)- Current Frequency: Three times a year ISSN: 1079-9060 Cancel/Invalid LCCN: sn 94006268 Notes: Published: [Atlanta, Ga.]: American Health Consultants, Each issue consists of a main entry section and three indexes. SERBIB/SERLOC merged record Subset of: BioScan. BioScan 0887-6207 (DLC) 88659731 (OCoLC)13302338 Subjects: Biotechnology industries-- Directories. Genetic engineering industry--Directories. Agricultural biotechnology--Directories. LC Classification: HD9999.B44 A37 Dewey Class No.: 338.7/63/0294 20

Agricultural & environmental biotechnology abstracts. Published/Created: Bethesda, MD: Cambridge Scientific Abstracts, c1993- Related Authors: Cambridge Scientific Abstracts, inc. Description: v.; 23 cm. Accompanied by an unnumbered, undated issue called "Sample issue." Vol. 1, no. 1- Current Frequency: Bimonthly Continues in part: Biotechnology research abstracts 0733-5709 (DLC) 85645979 (OCoLC)10692418 ISSN: 1063-1151 Cancel/Invalid LCCN: sn 92004782 Notes: SERBIB/SERLOC merged record Subjects: Agricultural biotechnology--Abstracts--Periodicals. Bioremediation--Abstracts--Periodicals. LC Classification: S494.5.B563 A363 Dewey Class No.: 660 20

Agricultural bioethics: implications of agricultural biotechnology / edited by Steven M. Gendel ... [et al.]. Edition Information: 1st ed. Published/Created: Ames: Iowa State University Press, 1990. Related Authors: Gendel, Steven M. Description: xxiv, 357 p.: ill.; 24 cm. ISBN: 081380129X Notes: Includes bibliographical references and index. Subjects: Agricultural biotechnology-- Moral and ethical aspects. LC Classification: S494.5.B563 A37 1990 Dewey Class No.: 174/.963 20

Agricultural biotechnology & the public good / edited by June Fessenden MacDonald. Published/Created: Ithaca, N.Y.: National Agricultural Biotechnology Council, c1994. Related Authors: MacDonald, June Fessenden. National Agricultural Biotechnology

Council (U.S.) NABC Meeting (6th: 1994: Michigan State University) Description: 213 p.: ill.; 23 cm. ISBN: 0963090755 Notes: Communicates the results of the sixth annual NABC meeting held at Michigan State University, East Lansing, Michigan on May 23-24, 1994. Includes bibliographical references. Subjects: Agricultural biotechnology-- Congresses. Agricultural biotechnology--Social aspects--Congresses. Public interest--Congresses. Series: NABC report; 6 LC Classification: S494.5.B563 A3723 1994

Agricultural biotechnology / edited by Arie Altman; international advisory editorial board, Rita R. Colwell ... [et al.]. Published/Created: New York: Marcel Dekker, c1998. Related Authors: Altman, A. Colwell, Rita R., 1934- Description: xiv, 770 p.: ill.; 26 cm. ISBN: 0824794397 (acid-free paper) Notes: Includes bibliographical references and index. Subjects: Agricultural biotechnology. Series: Books in soils, plants, and the environment LC Classification: S494.5.B563 A3715 1998 Dewey Class No.: 631 21

Agricultural biotechnology at the crossroads: biological, social & institutional concerns / edited by June Fessenden MacDonald. Published/Created: Ithaca, NY: National Agricultural Biotechnology Council, c1991. Related Authors: MacDonald, June Fessenden. University of California, Davis. Agricultural Issues Center. NABC Meeting (3rd: 1991: University of California-Davis) Description: 307 p.: ill.; 23 cm. Cancelled ISBN: 0963090740 Notes: Communicates the results of the Third

Annual NABC Meeting held in May 1991 at the University of California, Davis in cooperation with the University of California Agricultural Issues Center. Includes bibliographical references. Subjects: Agricultural biotechnology--Congresses. Agricultural biotechnology--United States--Congresses. Series: NABC report; 3 LC Classification: S494.5.B563 A366 1991 Dewey Class No.: 630 20

Agricultural biotechnology in developing countries: towards optimizing the benefits for the poor / edited by Matin Qaim, Anatole F. Krattiger, and Joachim von Braun. Published/Created: Boston: Kluwer Academic Publishers, c2000. Related Authors: Qaim, Matin. Krattiger, Anatole F. Von Braun, Joachim, 1950- Description: x, 433 p.: ill.; 25 cm. ISBN: 0792372301 (hb.: alk. paper) Notes: Papers from a conference held at the University of Bonn, Nov. 1999. Includes bibliographical references. Subjects: Agricultural biotechnology--Developing countries Congresses. LC Classification: S494.5.B563 A3727 2000 Dewey Class No.: 338.1/6/091724 21

Agricultural biotechnology in international development / edited by Catherine L. Ives and Bruce M. Bedford. Published/Created: New York: CABI Pub., c1998. Related Authors: Ives, C. (Catherine) Bedford, Bruce. Agricultural Biotechnology for Sustainable Productivity Project. Description: xiii, 354 p.: ill.; 24 cm. ISBN: 0851992781 (alk. paper) Notes: "This book is based on the proceedings of the 'Biotechnology for a better world' conference held at the Asilomar Conference Center, Pacific Grove,

California April 28-30, 1997. The conference was sponsored and coordinated by the Agricultural Biotechnology for Sustainable Productivity (ABSP) project ..." Acknowledgements. Includes bibliographical references and index. Subjects: Agricultural biotechnology--Developing countries Congresses. Series: Biotechnology in agriculture series; 21 LC Classification: S494.5.B563 A3728 1998 Dewey Class No.: 338.1/6 21

Agricultural biotechnology in the developing world. Published/Created: Rome: Food and Agriculture Organization of the United Nations, 1995. Related Authors: Food and Agriculture Organization of the United Nations. Description: vii, 162 p.; 30 cm. ISBN: 9251036268 Notes: Includes bibliographical references. Subjects: Agricultural biotechnology--Developing countries. Series: FAO research and technology paper, 1020-0541; 6 LC Classification: S494.5.B563 A373 1995 Dewey Class No.: 338.1/6 21

Agricultural biotechnology in the ICAR research institutes. Published/Created: New Delhi: Publications and Information Division, Indian Council of Agricultural Research, 1993. Related Authors: Indian Council of Agricultural Research. Publications and Information Division. Description: xii, 291 p.: col. ill.; 25 cm. Notes: Includes bibliographical references and index. Subjects: Agriculture--Research--India. Agricultural biotechnology--India Biotechnology Research--India. LC Classification: S471.I3 A53 1993

Agricultural biotechnology: a public conversation about risk / edited by June

Fessenden MacDonald.
Published/Created: Ithaca, NY: National Agricultural Biotechnology Council, c1993. Related Authors: MacDonald, June Fessenden. National Agricultural Biotechnology Council (U.S.) NABC Meeting (5th: 1993: Purdue University) Description: 135 p.: ill.; 23 cm. ISBN: 0963090739 Notes: Communicates the results of the fifth annual NABC meeting held in 1993 at Purdue University. Includes bibliographical references. Subjects: Agricultural biotechnology--Congresses. Agricultural biotechnology--United States--Congresses. Series: NABC report; 5 LC Classification: S494.5.B563 A3725 1993

Agricultural biotechnology: country case studies: a decade of development / edited by Gabrielle J. Persley, L. Reginald MacIntyre.
Published/Created: Wallingford: CABI Pub., c2002. Related Authors: Persley, Gabrielle J. (Gabrielle Josephine) MacIntyre, Reginald. Description: xix, 228 p.: ill., port.; 25 cm. ISBN: 0851988164 Notes: Includes bibliographical references and index. Subjects: Agricultural biotechnology-- Case studies. Series: Biotechnology in agriculture; no. 25 National Bib. No.: GB98-W5174

Agricultural biotechnology: issues and choices: information for decision makers / edited by Bill R. Baumgardt and Marshall A. Martin.
Published/Created: West Lafayette, Ind.: Purdue University Agricultural Experiment Station, c1991. Related Authors: Baumgardt, Bill R. Martin, Marshall A. Description: ix, 181 p.: ill.; 28 cm. ISBN: 0931682282 Notes: Includes bibliographical references.

Subjects: Agricultural biotechnology. LC Classification: S494.5.B563 A374 1991 Dewey Class No.: 338.1/6 20

Agricultural biotechnology: opportunities for international development / edited by Gabrielle J. Persley.
Published/Created: Wallingford, Oxon, UK: CAB International for the World Bank ... [et al.], 1993. Related Authors: Persley, G. J. World Bank. Description: xv, 495 p.: ill.; 26 cm. ISBN: 0851986439 Notes: Includes bibliographical references (p. 444-476) and index. Subjects: Biotechnology. Agricultural biotechnology. Biotechnology Series: Biotechnology in agriculture series; 2. Variant Series: Biotechnology in agriculture series, 0960-202X; no. 2 LC Classification: S494.5.B563 A376 1993 Dewey Class No.: 338.1/6 20

Agricultural biotechnology: prospects for the Third World / edited by John Farrington. Published/Created: London: Overseas Development Institute: Boulder, Colo.: Westview Press, c1989. Related Authors: Farrington, John. Overseas Development Institute (London, England) Description: 88 p.; 21 cm. ISBN: 0850031192 0813310822 (Westview): Notes: Includes bibliographical references (p. 66-70). Subjects: Agricultural biotechnology. Agricultural biotechnology--Developing countries. LC Classification: S494.5.B563 A38 1990 Dewey Class No.: 338.1/6/091724 20

Agricultural biotechnology: the next "green revolution"? / Agriculture and Rural Development Department, World Bank ... [et al.]. Published/Created: Washington, D.C.: World Bank, c1991. Related Authors: World Bank.

Agriculture and Rural Development Dept. Description: x, 51 p.; 28 cm. ISBN: 0821317415 Notes: Includes bibliographical references (p. 38-41). Subjects: Agricultural biotechnology--Developing countries Congresses. Agricultural innovations--Economic aspects--Developing countries--Congresses. Agricultural innovations--Social aspects--Developing countries--Congresses. Green Revolution--Congresses. Series: World Bank technical paper, 0253-7494; no. 133 LC Classification: HD1417 .A447 1990 Dewey Class No.: 338.1/61/091724 20

Agricultural commodities as industrial raw materials. Published/Created: Washington, D.C.: Congress of the U.S., Office of Technology Assessment: For sale by the Supt. of Docs., U.S. G.P.O., 1991. Related Authors: United States. Congress. Office of Technology Assessment. Description: vii, 109 p.; 26 cm. ISBN: 0160323606: Notes: Shipping list no.: 91-404-P. "May 1991"--P. [4] of cover. "OTA-F-476"--P. [4] of cover. 052-003-01237-5 Item 1070-M Includes bibliographical references. Subjects: Raw materials--United States. Farm produce--United States. Agricultural biotechnology--United States. LC Classification: HF1052 .A7 1991

Agriculture and intellectual property rights: economic, institutional, and implementation issues in biotechnology / edited by V. Santaniello ... [et al.]. Published/Created: Wallingford, Oxon, UK; New York: CABI Pub., c2000. Related Authors: Santaniello, V. Description: ix, 259 p.; 24 cm. ISBN: 0851994571 (alk. paper) Notes: Includes bibliographical references. Subjects: Germplasm resources, Plant--

Congresses. Agricultural biotechnology--Congresses. Intellectual property--Congresses. LC Classification: SB123.3 .A53 2000 Dewey Class No.: 631.5/23 21

Agriculture Sector Symposium (9th: 1989: World Bank) Innovation in resource management: proceedings of the Ninth Agriculture Sector Symposium / L. Richard Meyers, editor. Published/Created: Washington, D.C.: World Bank, c1989. Related Authors: Meyers, L. Richard. Description: vii, 280 p.; 28 cm. ISBN: 0821312820 Notes: Includes bibliographical references. Subjects: Agriculture--Congresses. Agricultural resources--Management--Congresses. Agriculture--Economic aspects--Congresses. Agricultural biotechnology--Congresses. Agricultural innovations--Congresses. Agricultural systems--Congresses. Natural resources--Management--Congresses. LC Classification: S401 .A475 1989 Dewey Class No.: 630 20

Agrobacterium protocols / edited by Kevan M.A. Gartland and Michael R. Davey. Published/Created: Totowa, N.J.: Humana Press, c1995. Related Authors: Gartland, K. M. A. Davey, M. R. (Michael Raymond), 1944- Description: xiv, 417 p.: ill.; 24 cm. ISBN: 0896033023 (acid-free paper) Notes: Includes bibliographical references and index. Subjects: Agrobacterium. Agricultural biotechnology. Series: Methods in molecular biology (Clifton, N.J.); v. 44. Variant Series: Methods in molecular biology; v. 44 LC Classification: QR82.R45 A37 1995 Dewey Class No.: 589.9/5 20

Ahunu, B. K. Livestock improvement: force or farce / B.K. Ahunu. Published/Created: Accra: Ghana Universities Press, 1988. Description: 32 p.; 21 cm. ISBN: 9964301553 Notes: "An inter-faculty lecture delivered at the University of Ghana on 3rd June 1987." Includes bibliographical references (p. 30-32). Subjects: Livestock--Breeding. Livestock--Genetics. Livestock improvement. Animal biotechnology. LC Classification: SF105 .A52 1988 Dewey Class No.: 636.08/2 20

An Assessment of the biotechnology/agribusiness industry: competition and the role of technology. Published/Created: Lancaster, Pa., USA: Technomic Pub. Co., [1985?] Related Authors: Technomic Publishing Company. Description: 152 p.: ill.; 28 cm. ISBN: 0877625999 Subjects: Agricultural industries--United States--Technological innovations. Agricultural innovations--Economic aspects--United States. Biotechnology industries--United States. Genetic engineering industry--United States. Crops--Genetic engineering--Economic aspects--United States. Livestock--Genetic engineering--Economic aspects--United States. Market surveys--United States. LC Classification: HD9005 .A87 1985 Dewey Class No.: 338.1/6 20

Angiosperm pollen and ovules / E. Ottaviano ... [et al., editors]. Published/Created: New York: Springer-Verlag, 1992. Related Authors: Ottaviano, Ercole. Description: xxvi, 465 p.: ill.; 24 cm. ISBN: 0387978887(New York) 3540978887 (Berlin) Notes: Includes bibliographical references and index. Subjects: Angiosperms--Pollen--Congresses. Pollen--Congresses. Ovules--Congresses. Plant biotechnology--Congresses. LC Classification: QK658 .A54 1992 Dewey Class No.: 582.13/04463 20

Animal biotechnology and the quality of meat production: papers presented at an OECD workshop held in Melle, Belgium, 7-9th November, 1990 / edited by L.O. Fiems and B.G. Cottyn and D.I. Demeyer. Published/Created: Amsterdam; New York: Elsevier; New York, NY, U.S.A.: Distributors for the U.S. and Canada, Elsevier Science Pub. Co., 1991. Related Authors: Fiems, L. O. Cottyn, B. G. Demeyer, D. I. Organisation for Economic Co-operation and Development. Description: xii, 218 p.: ill.; 25 cm. ISBN: 0444889302 (acid-free paper) Notes: "Organizers: Organisation for Economic Co-operation and Development (OECD)--National Institute for Animal Nutrition, Government Agricultural Research Centre-Ghent--Belgian Association for Meat Science and Technology (BAMST)." Includes bibliographical references. Subjects: Animal biotechnology--Congresses. Meat--Quality--Congresses. Series: Developments in animal and veterinary sciences; 25 LC Classification: SF140.B54 A56 1991 Dewey Class No.: 636.088/3 20

Animal biotechnology. Published/Created: New York, N.Y.: Marcel Dekker, c1990- Description: v.: ill.; 26 cm. Vol. 1, no. 1- Current Frequency: Semiannual ISSN: 1049-5398 Cancel/Invalid LCCN: sn 90005136 CODEN: ANBTEN Notes: Title from cover. SERBIB/SERLOC merged record Indexed entirely by: Index medicus 0019-3879 v9n1,1998- Indx'd selectively by: Bibliography of agriculture 0006-1530 Chemical abstracts 0009-2258 Additional Form Avail.: Also available to subscribers of OCLC FirstSearch Electronic Collections Online and Dekker via the World Wide Web. Subjects: Animal biotechnology--Periodicals. Animals, Domestic--genetics--periodicals. Animals, Transgenic--periodicals. Biotechnology--periodicals. LC Classification: SF140.B54 A55 Dewey Class No.: 636 20

Animal biotechnology: Comprehensive biotechnology, first supplement / volume editors, Lorne A. Babiuk and John P. Phillips; editor-in-chief, Murray Moo-Young. Edition Information: 1st ed. Published/Created: Oxford; New York: Pergamon Press, 1989. Related Authors: Babiuk, Lorne A. Phillips, John P. Moo-Young, Murray. Related Titles: [Comprehensive biotechnology. Description: xii, 260 p.: ill.; 28 cm. ISBN: 0080347304: Notes: Includes bibliographical references and index. Subjects: Veterinary medicine--Technological innovations. Animal biotechnology. Livestock--Breeding--Technological innovations. LC Classification: SF745 .A55 1989 Dewey Class No.: 636.089 20

Animal biotechnology: opportunities & challenges / edited by June Fessenden MacDonald. Published/Created: Ithaca, N.Y.: National Agricultural Biotechnology Council, 1992. Related Authors: MacDonald, June Fessenden. National Agricultural Biotechnology Council (U.S.) Description: 181 p.: ill.; 23 cm. ISBN: 0963090720: Notes: Papers from the Fourth Annual NABC Meeting held May 1992 in College Station, Tex. and hosted by Texas A&M University. Includes bibliographical references. Subjects: Animal biotechnology--Congresses. Series: NABC report; 4 LC Classification: SF140.B54 A555 1992

Animal cell bioreactors / edited by Chester S. Ho, Daniel I. C. Wang. Published/Created: Boston: Butterworth-Heinemann, c1991. Related Authors: Ho, Chester S., 1950. Wang, Daniel I-chyau, 1936- Description: xvi, 494 p.: ill.; 25 cm. ISBN: 0409901237 (alk. paper) Notes: Includes bibliographical references and index. Subjects: Animal cell biotechnology. Bioreactors. Series: Biotechnology (Reading, Mass.); 17. Variant Series: Biotechnology; 17 LC Classification: TP248.27.A53 A54 1991 Dewey Class No.: 660/.6 20

Animal cell biotechnology / edited by R.E. Spier, J.B. Griffiths. Published/Created: London; Orlando: Academic Press, 1985-<1988 Related Authors: Spier, R. (Raymond) Griffiths, J. B. Description: v. <1-3: ill.; 24 cm. ISBN: 0126575517 (v. 1) 0126575525 (v. 2) Notes: Includes bibliographies and index. Subjects: Cell culture. Biotechnology. LC Classification: QH585 .A58 1985 Dewey Class No.: 615/.36 19

Animal cell biotechnology: methods and protocols / edited by Nigel Jenkins. Published/Created: Totowa, N.J.: Humana Press, c1999. Related Authors: Jenkins, Nigel, 1954- Description: xiv, 302 p.: ill.; 24 cm. ISBN: 0896035476 (alk. paper) Notes: Includes bibliographical references and index. Subjects: Animal cell biotechnology. Cell culture. Series: Methods in biotechnology; 8 LC Classification: TP248.27.A53 A542 1999 Dewey Class

No.: 660.6 21

Animal cell electroporation and electrofusion protocols / edited by Jac A. Nickoloff. Cover Animal cell electroporation & electrofusion protocols Published/Created: Totowa, N.J.: Humana Press, c1995. Related Authors: Nickoloff, Jac A. Description: xx, 369 p.: ill.; 23 cm. ISBN: 089603304X (alk. paper) Notes: Includes bibliographical references and index. Subjects: Electroporation. Electrofusion. Animal cell biotechnology. Series: Methods in molecular biology (Clifton, N.J.); v. 48. Variant Series: Methods in molecular biology; v. 48 LC Classification: QH585.5.E48 A55 1995 Dewey Class No.: 591.87/028 20

Animal cell technology: challenges for the 21st century: proceedings of the joint international Association for Animal Cell Technology (JAACT) and the European Society for Animal Cell Technology (ESACT) 1998, Kyoto, Japan / edited by Kouji Ikura ... [et al.]. Published/Created: Dordrecht; Boston: Kluwer, c1999. Related Authors: Ikura, K‾oji, 1947- Japanese Association for Animal Cell Technology. European Society of Animal Cell Technology. Description: xiii, 452 p.: ill.; 25 cm. ISBN: 0792358058 (acid-free paper) Notes: "The second joint meeting of the Japanese Association for Animal Cell Technology (JAACT) and the European Society for Animal Cell Technology (ESACT) was held in Kyoto, Japan, July 26-30, 1998"--Pref. Includes bibliographical references and indexes. Subjects: Animal cell biotechnology--Congresses. LC Classification: TP248.27.A53 A543 1999 Dewey Class

No.: 660.6 21

Animal cell technology: developments towards the 21st century / edited by E.C. Beuvery, J.B. Griffiths and W.P. Zeijlemaker. Published/Created: Dordrecht; Boston: Kluwer Academic Publishers, c1995. Related Authors: Beuvery, E. C. Griffiths, J. B. Zeijlemaker, Wim P. Description: xxx, 1207 p.: ill.; 25 cm. ISBN: 0792337360 (hc: acid-free paper) Notes: Proceedings of the Veldhoven meeting. Includes bibliographical references and indexes. Subjects: Animal cell biotechnology--Congresses. LC Classification: TP248.27.A53 A545 1995 Dewey Class No.: 660/.6 20

Animal cell technology: developments, processes, and products / editors, R.E. Spier, J.B. Griffiths, C. MacDonald. Published/Created: Oxford; Boston: Butterworth-Heinemann, 1992. Related Authors: Spier, R. (Raymond) Griffiths, J. B. MacDonald, C. (Caroline) European Society of Animal Cell Technology. General Meeting (11th: Brighton, England) Description: xxvii, 756 p.: ill.; 25 cm. ISBN: 0750604212 Notes: "ESACT, European Society for Animal Cell Technology, the 11th Meeting." Includes bibliographical references and indexes. Subjects: Animal cell biotechnology--Congresses. LC Classification: TP248.27.A53 A544 1992 Dewey Class No.: 660/.6 20

Animal cell technology: from vaccines to genetic medicine / edited by Manuel J.T. Carrondo, Bryan Griffiths, and José L.P. Moreira. Published/Created: Dordrecht; Boston: Kluwer Academic Publishers, c1997. Related Authors: Carrondo, Manuel J. T. Griffiths, J. B. Moreira, José L. P. European Society of

Animal Cell Technology. General Meeting (14th: 1996: Vila Moura, Portugal) Description: xvi, 816 p.: ill.; 25 cm. ISBN: 0792343212 (hb: alk. paper) Notes: "Proceedings of the 14th Meeting of the ESACT, Vilamoura, Portugal, May 1996"--P. 1 of cover. Includes bibliographical references and indexes.

Atlantic Fisheries Technology Conference (34th: 1989: St. John's, Nfld.) Advances in fisheries technology and biotechnology for increased profitability: papers from the 34th Atlantic Fisheries Technological Conference and Seafood Biotechnology Workshop, August 27 to September 1, 1989, St. John's, NF, Canada / edited by Michael N. Voigt, J. Richard Botta. Published/Created: Lancaster: Technomic Pub. Co., c1990. Related Authors: Voigt, Michael N. Botta, J. Richard. Seafood Biotechnology Workshop (1989: St. John's, Nfld.) Description: xi, 566 p.: ill.; 28 cm. ISBN: 0877627851 Notes: Includes bibliographical references and index. Subjects: Fishery technology--Congresses. Biotechnology--Congresses. LC Classification: SH334.55 .A75 1989 Dewey Class No.: 639.2 20

Ayers, Jeanie H. Biotechnology in agriculture: advances in commercial livestock and plant production technology / Jeanie H. Ayers, James D. Greer. Published/Created: Menlo Park, Calif. (333 Ravenswood Ave., Menlo Park 94025-3476): SRI International, Business Intelligence Program, c1984. Related Authors: Greer, James D. Business Intelligence Program (SRI International) Description: 40 p.: ill.; 28 cm. Notes: Cover title. Subjects: Agricultural biotechnology--United States. Biotechnology industries--United States. Series: Report (Business Intelligence Program (SRI International)); no. 707. Variant Series: Report; no. 707 (fall 1984) LC Classification: S494.5.B563 A94 1984

Babu, Suresh Chandra. Economics of biotechnology: the case of biofertilizers in South Indian agriculture / by Suresh Candra Babu, J. Arne Hallam, Lehman B. Fletcher. Published/Created: Ames, Iowa, USA: Technology and Social Change Program, Iowa State University, c1988. Related Authors: Hallam, J. Arne. Fletcher, Lehman B. Description: 60 p.; 28 cm. ISBN: 0945271107: Notes: Includes bibliographical references (p. 59-60). Subjects: Fertilizer industry--India, South. Biotechnology industries--India, South. Agriculture--Economic aspects--India, South. Series: Studies in technology and social change series, 0896-1905; no. 7 LC Classification: HD9483.I43 S683 1988 Dewey Class No.: 338.4/766862/09548 20

Baculovirus expression systems and biopesticides / editors, Michael L. Shuler ... [et al.]. Published/Created: New York: Wiley-Liss, c1995. Related Authors: Shuler, Michael L., 1947- Description: x, 259 p.: ill.; 25 cm. ISBN: 0471065803 (acid-free paper) Notes: Includes bibliographical references and index. Subjects: Baculoviruses--Biotechnology. Insect cell biotechnology. Biological pest control agents. Baculoviridae--genetics. Gene Expression. Pest Control, Biological. Cells, Cultured. LC Classification: QR398.5 .B33 1995

Banana improvement: research challenges and opportunities / edited by Gabrielle J. Persley, Pamela George. Published/Created: Washington, D.C.: World Bank, c1996. Related Authors: Persley, G. J. George, Pamela, 1948- Description: ix, 47 p.; 28 cm. ISBN: 0821337408 Notes: Papers from a meeting held at Katholieke Universiteit Leuven, Feb. 1996. Includes bibliographical references (p. 47). Subjects: Bananas--Breeding--Congresses. Bananas--Biotechnology--Congresses. Bananas--Disease and pest resistance--Genetic aspects Congresses. Bananas--Research--Congresses. Crop improvement--Research--Congresses. Series: Environmentally sustainable development, agricultural research and extension group series. Banana Improvement Project report; no. 1 LC Classification: SB379.B2 B345 1996 Dewey Class No.: 634/.77223 20

Banana, breeding, and biotechnology: commodity advances through Banana Improvement Project Research, 1994-1998 / edited by Gabrielle J. Persley, Pamela George. Published/Created: Washington, D.C.: World Bank, c1999. Related Authors: Persley, G. J. George, Pamela, 1948- Description: x, 62 p.; 28 cm. ISBN: 0821344986 Notes: Includes bibliographical references. Subjects: Bananas--Breeding. Bananas--Biotechnology. Series: Environmentally sustainable development, agricultural research, and extension group series. Banana Improvement Project report; no. 2 LC Classification: SB379.B2 B347 1999 Dewey Class No.: 634/.77223 21

Bio technology: economic and wider impacts. Published/Created: Paris: Organisation for Economic Co-operation and Development; [Washington, D.C.: OECD Publications and Information Centre, distributor], 1989. Related Authors: Organisation for Economic Co-operation and Development. Related Titles: Biotechnology. Description: 111 p.: ill.; 23 cm. ISBN: 9264131965 Notes: Includes bibliographical references (p. 103-106). Subjects: Biotechnology industries. Agricultural innovations. LC Classification: HD9999.B442 B54 1989 Dewey Class No.: 338.4/76606 20

Bioactive compounds from plants. Published/Created: Chichester [England]; New York: Wiley, 1990. Projected Pub. Date: 1111 Related Authors: Chadwick, Derek. Marsh, Joan. Sath‾aban Wichai Chul‾aph‾on (Bangkok, Thailand) Symposium on Bioactive Compounds from Plants (1990: Bangkok, Thailand) Description: p. cm. ISBN: 0471926914 Notes: Editors: Derek Chadwick and Joan Marsh. Symposium on Bioactive Compounds from Plants, held in collaboration with the Chulabhorn Research Institute at the Royal Orchid Sheraton Hotel, Bangkok, Thailand, Feb. 20-22, 1990. "A Wiley-Interscience publication." Includes bibliographical references. Subjects: Plant bioactive compounds--Congresses. Plant biotechnology--Congresses. Series: Ciba Foundation symposium; 154 LC Classification: QK898.B54 B56 1990 Dewey Class No.: 660/.6 20

Biopesticide and biofertilizer markets: biotech weeds out chemical counterparts. Spine Biopesticide & biofertilizer markets Published/Created: Mountain View, CA: Market Intelligence, c1992. Related Authors: Frost & Sullivan. Market Intelligence

Research Corporation. Description: 1 v. (various pagings); 29 cm. ISBN: 1567533779 Notes: "Frost & Sullivan"--Cover. "246-52"--Cover. Subjects: Biological pest control agents industry--United States. Fertilizer industry--United States. Biotechnology industries--United States. Agricultural biotechnology--United States. Market surveys--United States. LC Classification: HD9999.B423 U62 1992

Biopolymers from polysaccharides and agroproteins / Richard A. Gross, editor, Carmen Scholtz, editor. Published/Created: Washington, DC: American Chemical Society, c2001. Related Authors: Gross, Richard A., 1957- Scholz, Carmen, 1963- Description: xiv, 426 p.: ill.; 24 cm. ISBN: 0841236453 (alk. paper) Notes: Includes bibliographical references and indexes. Subjects: Polymers--Biotechnology. Polysaccharides--Biotechnology. Plant proteins--Biotechnology. Biodegradation. Series: ACS symposium series; 786 LC Classification: TP248.65.P62 B547 2001 Dewey Class No.: 668.9 21

Bioresources and biotechnology: policy concerns for the Asian region / edited by Suman Sahai. Published/Created: New Delhi: Gene Campaign, c1999. Related Authors: Sahai, Suman. Description: x, 174 p.; 24 cm. ISBN: 8190100912 Summary: Contributed articles presented at a seminar. Subjects: Agricultural biotechnology--Asia--Congresses. Agricultural resources--Asia--Congresses. LC Classification: S494.5.B563 B527 1999 Dewey Class No.: 333.95/34/095 21

Bioseparation processes in foods / edited by Rakesh K. Singh, Syed S.H. Rizvi. Published/Created: New York: M. Dekker, c1995. Related Authors: Singh, R. K. (Rakesh K.) Rizvi, S. S. H., 1948- Description: viii, 469 p.: ill.; 24 cm. ISBN: 082479608X (acid-free) Notes: "The 18th symposium ... took place on June 24 and 25, 1994, in Atlanta"--Pref. Includes bibliographical references and index. Subjects: Food--Biotechnology. Separation (Technology) Series: IFT basic symposium series (Marcel Dekker, Inc.); 10. Variant Series: IFT basic symposium series; 10 LC Classification: TP248.65.F66 B49 1995 Dewey Class No.: 664/.02 20

Biosynthesis and manipulation of plant products / edited by Don Grierson. Edition Information: 1st ed. Published/Created: London; New York: Blackie Academic & Professional; New York: Chapman & Hall, 1993. Related Authors: Grierson, Donald. Description: x, 253 p.: ill.; 24 cm. ISBN: 0751400602 (acid-free text paper) Notes: Includes bibliographical references and index. Subjects: Plant biotechnology. Plant molecular biology. Botanical chemistry. Series: Plant biotechnology (Glasgow, Scotland); v. 3. Variant Series: Plant biotechnology; v. 3 LC Classification: TP248.27.P55 B57 1993 Dewey Class No.: 581.19/29 20

Biotechknowledgey: a report on the future of food: promise or peril? Published/Created: [Washington, DC]: Food and Allied Service Trades Dept., [1993?] Related Authors: AFL-CIO. Food and Allied Service Trades Dept. Description: 64 p.: ill.; 29 cm. Notes: "Food and Allied Service Trades Dept., AFL-CIO--F.A.S.T."--Prelim. p. Subjects: Biotechnology. Food--Biotechnology. Biotechnology industries--United States. Biotechnology--Law and legislation-- United States. LC Classification: TP248.2 .B378 1993 Dewey Class No.: 363.19/2 20

Biotechnology and the food industry / edited by P.L. Rogers and G.H. Fleet. Published/Created: New York: Gordon and Breach Science Publishers, c1989. Related Authors: Rogers, P. L. (Peter L.), 1940- Fleet, G. H. (Graham H.), 1946- University of New South Wales. Description: xvi, 298 p.: ill.; 24 cm. ISBN: 2881243541 Notes: Revised and updated contributions from a conference held at the University of New South Wales, Sydney, Feb. 26-27, 1987. Includes bibliographies and index. Subjects: Food--Biotechnology-- Congresses. LC Classification: TP248.65.F66 B57 1989 Dewey Class No.: 664 20

Biotechnology and the improvement of forage legumes / edited by B.D. McKersie and D.C.W. Brown. Published/Created: Wallingford, Oxon, UK; New York, NY, USA: CAB International, 1997. Related Authors: McKersie, Bryan D. Brown, D. C. W. (Daniel C. W.) Description: ix, 444 p.: ill.; 26 cm. ISBN: 0851991092 (alk. paper) Notes: Includes bibliographical references and index. Subjects: Legumes--Biotechnology. Forage plants--Biotechnology. Crop improvement. LC Classification: SB203 .B56 1997 Dewey Class No.: 633.3/0423 21

Biotechnology and the new agricultural revolution / edited by Joseph J. Molnar and Henry Kinnucan. Published/Created: Boulder, Colo.: Westview Press for the American Association for the Advancement of Science, Washington, D.C., 1989. Related Authors: Molnar, Joseph J. Kinnucan, Henry W. Description: xv, 288 p.; 23 cm. ISBN: 081337667X (alk. paper) Notes: Includes bibliographies and indexes. Subjects: Agricultural biotechnology. Agricultural innovations. Series: AAAS selected symposium; 108 LC Classification: S494.5.B563 B54 1989 Dewey Class No.: 630 19

Biotechnology and the U.S. food industry: a study / prepared by the Office of Planning and Evaluation, and the Center for Food Safety and Applied Nutrition, U.S. F.D.A. Published/Created: Lancaster, Pa., USA: Technomic Pub. Co., [1988] Related Authors: United States. Food and Drug Administration. Office of Planning and Evaluation. Center for Food Safety and Applied Nutrition (U.S.) Related Titles: Biotechnology and the United States food industry. Description: vi, 146 p. :ill.; 28 cm. ISBN: 0877626413 Notes: Includes bibliographical references (p. 145-146). Subjects: Food-- Biotechnology. Food industry and trade- -United States--Statistics. LC Classification: TP248.65.F66 B59 1988 Dewey Class No.: 664 20

Biotechnology applications in agriculture in Asia and the Pacific: report of an APO study meeting 18th-28th January, 1994, Tokyo, Japan. Published/Created: Tokyo: Asian Productivity Organization, 1994. Related Authors: Asian Productivity Organization. APO Study Meeting on Biotechnology Applications in Agriculture (1994: Tokyo, Japan) Description: 329 p.: ill.; 26 cm. ISBN: 9283321529 Notes: Includes country reports from Bangladesh, Republic of China, Fiji,

Hong Kong, India, Indonesia, Iran, Republic of Korea, Malaysia, Mongolia, Nepal, Pakistan, Philippines, Sri Lanka, and Thailand. "Report of the APO Study Meeting on Biotechnology Applications in Agriculture held in Japan from 18th to 28th January, 1994."--T.p. verso. "STM-05-94"--T.p. verso. Includes bibliographical references. Subjects: Biotechnology--Asia--Congresses. Biotechnology--Pacific Area--Congresses. Biotechnology industries--Asia--Congresses. Biotechnology industries--Pacific Area--Congresses. LC Classification: TP248.195.A78 B55 1994

Biotechnology applications in beverage production / edited by C. Cantarelli and G. Lanzarini. Published/Created: London; New York: Elsevier Applied Science; New York, NY, USA: Sole distributor in the USA and Canada, Elsevier Science Pub., c1989. Related Authors: Cantarelli, C. (Corrado) Lanzarini, G. Description: x, 257 p.: ill.; 23 cm. ISBN: 1851663282 Notes: Includes bibliographical references and index. Subjects: Beverages--Congresses. Biotechnology--Industrial applications--Congresses. Series: Elsevier applied food science series LC Classification: TP501 .B56 1989 Dewey Class No.: 663 19

Biotechnology challenges for the flavor and food industry / edited by Robert C. Lindsay and Brian J. Willis. Published/Created: London [England]; New York: Elsevier Applied Science, c1989. Related Authors: Lindsay, Robert C. (Robert Clarence), 1936- Willis, Brian J. Quest International (Organization: Naarden, Netherlands) Description: x, 170 p.: ill.; 24 cm. ISBN: 185166405X Notes:

"Proceedings of an international symposium organized by Quest International at Williamsburg Lodge in Williamsburg, Virginia, USA, 2-5 October, 1988." Includes bibliographical references. Subjects: Flavor--Biotechnology--Congresses. Food--Biotechnology--Congresses. LC Classification: TP418 .B57 1989 Dewey Class No.: 664/.07 20

Biotechnology in animal husbandry / edited by R. Renaville and A. Burny. Published/Created: Dordrecht; Boston: Kluwer Academic Publishers, c2001. Related Authors: Renaville, R. Burny, A. Description: 360 p.: ill.; 25 cm. ISBN: 0792368517 (hardcover) Notes: "Volume 5." Includes bibliographical references and index. Subjects: Animal biotechnology. Livestock--Genetic engineering. Series: Focus on biotechnology; v. 5 LC Classification: SF140.B54 B58 2001 Dewey Class No.: 636.08/21 21

Biotechnology in food processing / edited by Susan K. Harlander and Theodore P. Labuza. Published/Created: Park Ridge, N.J., U.S.A.: Noyes Publications, c1986. Related Authors: Harlander, Susan K. LaBuza, Theodore Peter, 1940- University of Minnesota. Dept. of Food Science and Nutrition. Description: xxvi, 323 p.: ill.; 25 cm. ISBN: 081551073X: Notes: Proceedings of a symposium held Oct. 7-9, 1985, at the University of Minnesota, sponsored by Dept. of Food Science and Nutrition at the University of Minnesota. Includes bibliographies and index. Subjects: Biotechnology. Food industry and trade. LC Classification: TP248.2 .B5537 1986 Dewey Class No.: 664 19

Biotechnology in horticultural and plantation crops / [editors] K.L. Chadha, P.N. Ravindran, Leela Sahijram. Published/Created: New Delhi: Malhotra Pub. House, c2000. Related Authors: Chadha, K. L. Ravindran, P. N. Sahijram, Leela. Description: xv, 836 p.: col. ill.; 25 cm. ISBN: 8185048428 Summary: Contributed articles. Notes: Includes bibliographical references and index. Subjects: Plant biotechnology. LC Classification: SB106.B56 .B556 2000 Dewey Class No.: 630 21

Biotechnology in plant disease control / edited by Ilan Chet. Published/Created: New York: Wiley-Liss, c1993. Related Authors: Chet, Ilan, 1939- Description: xvi, 373 p.: ill.; 24 cm. ISBN: 0471560847 (acid-free paper) Notes: Includes bibliographical references and index. Subjects: Plants--Disease and pest resistance--Genetic aspects. Phytopathogenic microorganisms-- Biological control. Plant diseases. Agricultural biotechnology. Series: Wiley series in ecological and applied microbiology LC Classification: SB750 .B56 1993 Dewey Class No.: 632/.3 20

Biotechnology in plant science: relevance to agriculture in the eighties / edited by Milton Zaitlin, Peter Day, and Alexander Hollaender; technical editor, Claire M. Wilson. Published/Created: Orlando [Fla.]: Academic Press, 1985. Related Authors: Zaitlin, Milton. Day, Peter R., 1928- Hollaender, Alexander, 1898- Description: xviii, 364 p.: ill.; 24 cm. ISBN: 0127753109 (alk. paper) Notes: Based on the proceedings of a symposium held June 23-27, 1985 at Cornell University. Includes bibliographies and index. Subjects: Agricultural biotechnology-- Congresses. Plant biotechnology--

Congresses. LC Classification: S494.5.B563 B58 1985 Dewey Class No.: 631 19

Biotechnology in the feed industry: proceedings of Alltech's fourth annual symposium / edited by T.P. Lyons. Published/Created: Nicholasville, Ky.: Alltech Technical Publications, c1988. Related Authors: Lyons, T. P. Alltech Biotechnology Center. Description: 362 p., [29] p. of plates: ill.; 24 cm. Notes: "Alltech Biotechnology Center"--Cover. Includes bibliographical references. Subjects: Animal nutrition--Congresses. Animal biotechnology--Congresses. Feeds--Congresses. Feed additives-- Congresses. Feed industry--Congresses. Feed additive industry--Congresses. LC Classification: SF95 .B54 1988

Biotechnology in the feed industry: proceedings of Alltech's tenth annual symposium / edited by T.P. Lyons and K.A. Jacques. Published/Created: Loughborough, England: Nottingham University Press, c1994. Related Authors: Lyons, T. P. Jacques, Kathryn Ann. Alltech Biotechnology Center. Description: viii, 344 p.: ill.; 24 cm. ISBN: 1897676514 Notes: Includes bibliographical references and index. Subjects: Animal nutrition--Congresses. Animal biotechnology--Congresses. Feeds--Congresses. Feed additives-- Congresses. Feed industry--Congresses Feed additive industry--Congresses. LC Classification: SF94.6 .B56 1994 Dewey Class No.: 636.08/52 21

Biotechnology of ornamental plants / edited by R.L. Geneve, J.E. Preece and S.A. Merkle. Published/Created: Wallingford, Oxon, UK; New York, NY, USA: CAB International, c1997. Related Authors: Geneve, R. L. Preece,

John E. Merkle, Scott Arthur, 1953-
Description: x, 402 p.: ill.; 25 cm.
ISBN: 0851991106 Notes: Includes
bibliographical references and index.
Subjects: Plants, Ornamental--
Biotechnology. Series: Biotechnology in
agriculture series; no. 16 LC
Classification: SB404.9 .B56 1997
Dewey Class No.: 635.9/15233 21

Biotechnology of perennial fruit crops /
edited by F.A. Hammerschlag & R.E.
Litz. Published/Created: Wallingford,
Oxon, UK: C.A.B. International, c1992.
Related Authors: Hammerschlag, F. A.
Litz, Richard E. Description: xxi, 550
p.: ill.; 26 cm. ISBN: 0851987087
Notes: Includes bibliographical
references and index. Subjects: Fruit--
Propagation. Fruit--Biotechnology.
Series: Biotechnology in agriculture
series; 8 LC Classification: SB359.3
.B55 1992 Dewey Class No.: 634/.043
21

Biotechnology of plant fats and oils / edited
by James Rattray. Published/Created:
Champaign, Ill.: American Oil
Chemists' Society, c1991. Related
Authors: Rattray, James. Description:
iv, 176 p.: ill.; 24 cm. ISBN:
0935315330 Notes: Includes
bibliographical referneces. Subjects:
Plant lipids--Biotechnology. LC
Classification: TP248.65 .P53B56 1991
Dewey Class No.: 664/.3 20

Biotechnology of plant-microbe interactions
/ [edited by] James P. Nakas, Charles
Hagedorn. Published/Created: New
York: McGraw-Hill, c1990. Related
Authors: Nakas, James P. Hagedorn,
Charles. Description: xi, 348 p.: ill.; 24
cm. ISBN: 0070458677: Notes:
Includes bibliographical references.
Subjects: Plant-microbe relationships--

Molecular aspects. Microbial
biotechnology. LC Classification:
QR351 .B56 1990 Dewey Class No.:
576/.15 20

Biotechnology of plants and microorganisms
/ edited by O.J. Crocomo ... [et al.].
Published/Created: Columbus: Ohio
State University Press, c1986. Related
Authors: Crocomo, Otto J. Description:
xiv, 473 p.; 26 cm. ISBN: 0814203752:
Notes: Includes bibliographies and
index. Subjects: Genetic engineering.
Plant genetic engineering. Microbial
genetics. Biotechnology. LC
Classification: TP248.6 .B557 1986
Dewey Class No.: 660/.6 19

Biotechnology, agriculture and the
developing world: the distributional
implications of technological change /
edited by Timothy Swanson.
Published/Created: Northampton, MA:
Edward Elgar Pub., 2002. Projected
Pub. Date: 0203 Related Authors:
Swanson, Timothy M. Description: p.
cm. ISBN: 1840646799 Notes: Includes
bibliographical references (p.).
Subjects: Agricultural biotechnology--
Developing countries. Agriculture--
Technology transfer--Developing
countries. Agrobiodiversity--
Developing countries. LC
Classification: S494.5.B563 B536 2002
Dewey Class No.: 631.5/233/091724 21

Biotechnology, agriculture, and food.
Published/Created: Paris: Organisation
for Economic Co-operation and
Development, 1992. Related Authors:
Organisation for Economic Co-
operation and Development.
Description: 219 p.; 23 cm. ISBN:
9264137254 Notes: Includes
bibliographical references. Subjects:
Food--Biotechnology. Agricultural

biotechnology. Biotechnology--
Government policy. LC Classification:
TP248.65.F66 B545 1992 Dewey Class
No.: 338.1/6 20

Biotechnology, biosafety, and biodiversity:
scientific and ethical issues for
sustainable development / editors,
Sivramish Shantharam, Jane F.
Montgomery. Published/Created:
Enfield, N.H.: Science Publishers,
c1999. Related Authors: Shantharam,
Sivramiah, 1951- Montgomery, Jane F.
Satellite Symposium on Biotechnology
and Biodiversity (1996: New Delhi,
India) Description: xiv, 237 p.: ill.; 25
cm. ISBN: 1578080185 Notes:
Proceedings of the Satellite Symposium
on Biotechnology and Biodiversity, held
on November 15 and 16, 1996, at the
Indian Agricultural Research Institute
(IARI), New Delhi, India. Includes
bibliographical references and index.
Subjects: Agricultural biotechnology--
Congresses. Agrobiodiversity--
Congresses. LC Classification:
S494.5.B563 B545 1999 Dewey Class
No.: 631.5/233 21

Calder, Nigel. The green machines / Nigel
Calder. Published/Created: New York:
Putnam, c1986. Description: 205 p.; 23
cm. ISBN: 0399131760 Notes:
Bibliography: p. 201-205. Subjects:
Biotechnology. Agricultural
biotechnology. LC Classification:
TP248.2 .C35 1986 Dewey Class No.:
303.4/9 19

Cereal biotechnology / edited by Peter C.
Morris, James H. Bryce.
Published/Created: Boca Raton, FL:
CRC Press/Woodhead Pub., 2000.
Related Authors: Morris, Peter C.
Bryce, James H. Description: x, 252 p.:
ill.; 25 cm. ISBN: 0849308992 (alk.

paper) Notes: Includes bibliographical
references. Subjects: Grain--
Biotechnology. LC Classification:
SB189 .C39 2000 Dewey Class No.:
633.1/04233 21

Cereals: novel uses and processes / edited by
Grant M. Campbell, Colin Webb, and
Stephen L. McKee. Published/Created:
New York: Plenum Press, c1997.
Related Authors: Campbell, Grant M.
Webb, Colin. McKee, Stephen L.
Description: xvi, 289 p.: ill.; 26 cm.
ISBN: 0306455838 Notes: "Proceedings
of an international conference on
Cereals: Novel Uses and Precesses, held
June 4-6, 1996, in Manchester, United
Kingdom"--T.p. verso. Includes
bibliographical references and index.
Subjects: Grain--Biotechnology. LC
Classification: TP248.27.P55 C47 1997
Dewey Class No.: 620.1/17 21

Chemicals via higher plant bioengineering /
edited by Fereidoon Shahidi ... [et al.].
Published/Created: New York: Kluwer
Academic/Plenum, c1999. Related
Authors: Shahidi, Fereidoon, 1951-
Chemical Congress of North America
(5th: 1997: Cancun, Mexico)
Description: viii, 280 p.: ill.; 26 cm.
ISBN: 030646117X Notes: "Based on
proceedings of a symposium on
Chemicals via Higher Plant
Engineering, held at the 5th North
American Chemical Congress,
November 11-15, 1997, in Cancun,
Mexico"--T.p. verso. Includes
bibliographical references and index.
Subjects: Plant biotechnology--
Congresses. Phytochemicals--
Congresses. Series: Advances in
experimental medicine and biology; v.
464 LC Classification: SB106.B56 C48
1999 Dewey Class No.: 660/.6 21

CTA/FAO Symposium (1989: Luxembourg, Luxembourg) Plant biotechnologies for developing countries: CTA/FAO Symposium, Luxembourg, 26-30 June 1989: conclusions and recommendations. Published/Created: Ede-Wageningen, The Netherlands: Technical Centre for Agricultural and Rural Cooperation; [Paris, France]: UN Food and Agriculture Organization, [1989?] Related Authors: Technical Centre for Agricultural and Rural Cooperation (Ede, Netherlands) Food and Agriculture Organization of the United Nations. Description: 50 p.; 30 cm. Subjects: Plant biotechnology--Congresses. Plant biotechnology--Developing countries--Congresses. LC Classification: SB106.B56 C78 1989 Dewey Class No.: 631.5/23 20

Cummins, Ronnie. Genetically engineered food: a self-defense guide for consumers / by Ronnie Cummins and Ben Lilliston; foreword by Andrew Kimbrell. Published/Created: New York; Marlowe, c2000. Related Authors: Lilliston, Ben. Description: xvi, 208 p.; 21 cm. ISBN: 1569246351 Notes: Includes bibliographical references (p. 201-208). Subjects: Food--Biotechnology--Popular works. Crops--Genetic engineering--Popular works. LC Classification: TP248.65.F66 C85 2000 Dewey Class No.: 363.19/2 21

Current plant science and biotechnology in agriculture. Published/Created: Dordrecht; Boston: M. Nijhoff Hingham, MA, USA: Distributors for the U.S. and Canada, Kluwer Academic Publishers, 1985- Description: v.: ill.; 25 cm. 1- Current Frequency: Irregular ISSN: 0924-1949 Cancel/Invalid LCCN: sn 87024299 CODEN: CPBAE2 Notes: Published: Dordrecht; Boston:

Kluwer Academic Publishers, <1992- Indx'd selectively by: Chemical abstracts 0009-2258 GeoRef 0197-7482 Subjects: Agricultural biotechnology. Plant genetic engineering. Plant tissue culture. LC Classification: CLASSED SEPARATELY NAL

Designer oil crops: breeding, processing, and biotechnology / edited by Denis J. Murphy. Published/Created: Weinheim; New York: VCH, c1994. Related Authors: Murphy, Denis J. Description: xvi, 317 p.: ill.; 25 cm. ISBN: 3527300406 (Weinheim: acid-free paper) 1560818271 (New York: acid-free paper) Notes: Includes bibliographical references and index. Subjects: Oilseed plants--Breeding. Oilseed plants--Biotechnology. Oilseed products. LC Classification: SB298 .D47 1994 Dewey Class No.: 665/.2 20

Designing value-added soybeans for markets of the future / edited by Richard F. Wilson. Published/Created: Champaign, Ill.: American Oil Chemists' Society, c1991. Related Authors: Wilson, Richard F., 1947- American Oil Chemists' Society. Meeting (81st: 1990: Baltimore, Md.) Description: vi, 135 p.: ill., map; 24 cm. ISBN: 0935315373 Notes: Based on presentations given at the 81st annual American Oil Chemists Society Meeting held in Baltimore, Md., Apr. 22-29, 1990. Includes bibliographical references. Subjects: Soybean--Germplasm resources--United States--Congresses. Soybean--Biotechnology--Economic aspects--United States Congresses. Soybean--Composition--Congresses. Soybean industry--Congresses. LC Classification: SB205.S7 D47 1991 Dewey Class No.:

633.3/42 20

Directory of agricultural and industrial biotechnology. Published/Created: Harare: Zimbabwe Biotechnology Advisory Committee, Secretariat, [1996] Related Authors: Zimbabwe Biotechnology Advisory Committee. Secretariat. Description: 76 p.; 21 cm. Notes: Cover title. "July, 1996"--P. 1. Includes index. Subjects: Agricultural biotechnology--Zimbabwe--Directories. Agricultural biotechnology--Research--Zimbabwe Directories. Biotechnology--Zimbabwe--Directories. Biotechnology--Research--Zimbabwe--Directories. LC Classification: S494.5.B563 D57 1996 Dewey Class No.: 631.5/233/0256891 21

Doyle, Jack, 1947- Altered harvest: agriculture, genetics, and the fate of the world's food supply / by Jack Doyle. Published/Created: New York, NY: Penguin Books, 1986, c1985. Description: xix, 502 p.; 21 cm. ISBN: 0140096965 (pbk.) Notes: Includes index. Bibliography: p. 388-431. Subjects: Food industry and trade--Technological innovations Government policy--United States. Agricultural innovations--Government policy--United States. Genetic engineering industry--Government policy--United States. Biotechnology industries--Government policy--United States. Seed industry and trade--Technological innovations Government policy--United States. Agricultural chemicals industry--Government policy--United States. Food adulteration and inspection--Government policy United States. LC Classification: HD9006 .D65 1986 Dewey Class No.: 338.1/9/73 19

Doyle, Jack, 1947- Altered harvest: agriculture, genetics, and the fate of the world's food supply / by Jack Doyle. Published/Created: New York, N.Y., U.S.A.: Viking, 1985. Description: xix, 502 p.; 24 cm. ISBN: 067011524X: Notes: Includes index. Bibliography: p. 388-431. Subjects: Food industry and trade--Technological innovations Government policy--United States. Agricultural innovations--Government policy--United States. Genetic engineering industry--Government policy--United States. Biotechnology industries--Government policy--United States. Seed industry and trade--Technological innovations Government policy--United States. Agricultural chemicals industry--Government policy--United States. Food adulteration and inspection--Government policy United States. LC Classification: HD9006 .D65 1985 Dewey Class No.: 338.1/9/73 19

E.B.C.-Symposium Plant Biotechnology (1989: Helsinki, Finland) E.B.C.-Symposium Plant Biotechnology, Helsinki, Finland, November 1989. Published/Created: Nürnberg: H. Carl Getränke-Fachverlag, c1990. Related Authors: European Brewery Convention. Related Titles: EBC-Symposium Plant Biotechnology, Helsinki, Finland, November 1989. Description: x, 160 p.: ill.; 24 cm. ISBN: 3418007236 Notes: Includes bibliographical references. Subjects: Barley--Biotechnology--Congresses. Hops--Biotechnology--Congresses. Brewing--Congresses. Series: Monograph (European Brewery Convention); 15. Variant Series: Monograph / European Brewery Convention; 15 LC Classification: SB191.B2 E2 1989 Dewey Class No.:

633.1/6 20

Engineering plants for commercial products and applications / edited by Glenn B. Collins and Robert J. Shepherd. Published/Created: New York: New York Academy of Sciences, 1996. Related Authors: Collins, Glenn B. Shepherd, Robert J. University of Kentucky. Tobacco and Health Research Institute. Description: 183 p.: ill.; 24 cm. ISBN: 1573310468(cloth: alk. paper) 1573310476(pbk.: alk. paper) Notes: This volume contains the scientific presentations of a symposium sponsored by the University of Kentucky Tobacco and Health Research Institute and held in Lexington, Ky., October 1-4, 1995. Includes bibliographical references and indexes. Subjects: Plant biotechnology--Congresses. Plant genetic engineering--Congresses. Series: Annals of the New York Academy of Sciences, 0077-8923; v. 792 LC Classification: Q11 .N5 vol. 792 Dewey Class No.: 660/.6 20

Engineering the farm: the social and ethical aspects of agricultural biotechnology / [edited by] Britt Bailey, Marc Lappe. Published/Created: Washington, D.C.: Island Press, 2002. Projected Pub. Date: 0206 Related Authors: Bailey, Britt. Lappé, Marc. Description: p. cm. ISBN: 1559639466 (hardcover: alk. paper) 1559639474 (pbk.: alk. paper) Notes: Includes bibliographical references. Subjects: Genetic engineering--Social aspects. Agricultural biotechnology--Social aspects. Genetic engineering--Moral and ethical aspects. Agricultural biotechnology--Moral and ethical aspects. LC Classification: S494.5.G44 E54 2002 Dewey Class No.: 174/.96315233 21

Enzymes for carbohydrate engineering / edited by Kwan-Hwa Park, John F. Robyt, Yang-Do Choi. Published/Created: Amsterdam [Netherlands]; New York: Elsevier, 1996. Related Authors: Park, Kwan-Hwa. Robyt, John F., 1935- Choi, Yang-Do. Description: vii, 215 p.: ill.; 25 cm. ISBN: 0444824081 (alk. paper) Notes: Includes bibliographical references and index. Subjects: Enzymes--Biotechnology. Carbohydrates--Biotechnology. Agricultural biotechnology. Series: Progress in biotechnology; 12 LC Classification: TP248.65.E59 E5916 1996 Dewey Class No.: 660/.634 20

Fernandez-Cornejo, Jorge. Genetically engineered crops for pest management in U.S. Agriculture: farm-level effects / Jorge Fernandez-Cornejo and William D. McBride; with contributions from Cassandra Klotz-Ingram ... [et al.] Published/Created: Washington, D.C. (1800 M St., NW, Washington 20036-5831): U.S. Dept. of Agriculture, Economic Research Service, [2000] Related Authors: McBride, William D. Description: iii, 20 p.: ill., map.; 28 cm. Notes: Cover title. "April 2000." Includes bibliographical references (p. 19-20). Subjects: Plants--Disease and pest resistance--United States Genetic aspects. Crops--Genetic engineering--United States. Agricultural biotechnology--United States. Series: Agricultural economic report; no. 786 LC Classification: HD1751 .A91854 no. 786 SB750 Dewey Class No.: 632/.9 21

Food and new biotechnology: novelty, safety, and control aspects of foods made by new biotechnology. Edition Information: 2. ed. Published/Created: Copenhagen: Nordic Council of

Ministers, 1992. Related Authors: Nordic Council of Ministers. Description: 247 p.: ill.; 25 cm. ISBN: 8773035734 (Denmark) Cancelled ISBN: 9179963399 (Sweden) Notes: Text in English; summary also in Danish and Swedish. Includes bibliographical references. Subjects: Food--Biotechnology. Series: Nord (Series); 1991:18 Variant Series: Nord; 1991:18 LC Classification: TP248.65.F66 F64 1992 Dewey Class No.: 664 21

Food biotechnology / edited by Stanislaw Bielecki, Johannes Tramper, Jacek Polak. Edition Information: 1st ed. Published/Created: Amsterdam; New York: Elsevier, 2000. Related Authors: Bielecki, Stanislaw. Tramper, J., 1949- Polak, Jacek. Description: xv, 430 p.: ill.; 25 cm. ISBN: 0444505199 (acid-free paper) Notes: "Proceedings of an international symposium organized by the Institute of Technical Biochemistry, Technical University of Lodz, Poland ... May 9-12, 1999." Includes bibliographical references and index. Subjects: Food--Biotechnology--Congresses. Series: Progress in biotechnology; 17 LC Classification: TP248.65.F66 F648 2000 Dewey Class No.: 664 21

Food biotechnology. Published/Created: London; New York: Elsevier Applied Science, 1987- Description: v.: ill.; 23 cm. 1- ISSN: 0952-357X Cancel/Invalid LCCN: sn 88024013 Notes: SERBIB/SERLOC merged record Indexed by: Bibliography of Agriculture 0006-1530 v. 2, 1988. Subjects: Food--Biotechnology--Periodicals. Food--Composition--Periodicals. Food--Analysis--Periodicals. LC Classification: TP248.65.F66 F65

Dewey Class No.: 664 20

Food biotechnology. Published/Created: New York, N.Y.: Dekker, c1987- Description: v.: ill.; 26 cm. Vol. 1, no. 1- Current Frequency: Two no. a year ISSN: 0890-5436 Cancel/Invalid LCCN: sn 86011621 CODEN: FBIOEE Notes: SERBIB/SERLOC merged record Indexed entirely by: Bibliography of agriculture 0006-1530 Indx'd selectively by: Biological abstracts 0006-3169 1987- Additional Form Avail.: Issued also in microform by Research Publications. Also available via the World Wide Web. Subjects: Food--Biotechnology--Periodicals. Biotechnology--periodicals. Food Technology--periodicals. Food-Processing Industry--periodicals. LC Classification: TP248.65.F66 F66 Dewey Class No.: 664 19

Food biotechnology: microorganisms / edited by Y.H. Hui, George G. Khachatourians. Published/Created: New York: VCH, c1995. Related Authors: Hui, Y. H. (Yiu H.) Khachatourians, George G., 1940- Description: xvi, 937 p.: ill.; 26 cm. ISBN: 1560815655 (acid-free paper) Notes: Includes bibliographical references and index. Subjects: Food--Biotechnology. Food--Microbiology. Series: Food science and technology (VCH Publishers) Variant Series: Food science and technology LC Classification: TP248.65.F66 F685 1995 Dewey Class No.: 664/.024 20

Food flavor and safety: molecular analysis and design / A.M. Spanier, editor, H. Okai, editor, M. Tamura, editor. Published/Created: Washington, DC: American Chemical Society, 1993. Related Authors: Spanier, A. M. (Arthur

M.), 1948- Okai, H. (Hideo), 1938-
Tamura, M. (Masahiro), 1956-
American Chemical Society. Division
of Agricultural and Food Chemistry.
American Chemical Society. Meeting
(203rd: 1992: San Francisco, Calif.)
Description: xiv, 352 p.: ill.; 24 cm.
ISBN: 0841226652 (alk. paper) Notes:
"Developed from a symposium
sponsored by the Division of
Agricultural and Food Chemistry at the
203rd national meeting of the American
Chemical Society, San Francisco,
California, April 5-10, 1992." Includes
bibliographical references and indexes.
Subjects: Food--Biotechnology--
Congresses. Series: ACS symposium
series; 528 LC Classification:
TP248.65.F66 F69 1993 Dewey Class
No.: 664 20

Food lipids: chemistry, nutrition, and
biochemistry / edited by Casimir C.
Akoh, David B. Min. Edition
Information: 2nd ed.
Published/Created: New York: M.
Dekker, c2002. Projected Pub. Date:
0205 Related Authors: Akoh, Casimir
C., 1955- Min, David B. Description: p.
cm. ISBN: 0824707494 (alk. paper)
Subjects: Lipids. Lipids in human
nutrition. Lipids--Biotechnology.
Lipids--Metabolism. Series: Food
science and technology (Marcel Dekker,
Inc.); 117. Variant Series: Food science
and technology; 117 LC Classification:
QP751 .F647 2002 Dewey Class No.:
612.3/97 21

Food lipids: chemistry, nutrition, and
biotechnology / edited by Casimir C.
Akoh, David B. Min.
Published/Created: New York: Marcel
Dekker, c1998. Related Authors: Akoh,
Casimir C., 1955- Min, David B.
Description: xi, 816 p.: ill.; 27 cm.

ISBN: 0824799852 (alk. paper) Notes:
Includes bibliographical references and
index. Subjects: Lipids. Lipids in human
nutrition. Lipids--Biotechnology.
Lipids--Metabolism. Series: Food
science and technology (Marcel Dekker,
Inc.); 88. Variant Series: Food science
and technology; 88 LC Classification:
QP751 .F647 1998 Dewey Class No.:
612.3/97 21

Food safety evaluation. Published/Created:
Paris: Organisation for Economic Co-
operation and Development;
Washington, D.C.: OECD Publications
and Information Center [distributor],
c1996. Related Authors: Organisation
for Economic Co-operation and
Development. Environment Directorate.
Organisation for Economic Co-
operation and Development. Directorate
for Science, Technology and Industry.
OECD Workshop on Food Safety
Evaluation (1994: Oxford, England).
Description: 180 p.: ill.; 27 cm. ISBN:
9264148671 Notes: Papers presented at
the OECD Workshop on Food Safety
Evaluation, held in Oxford, England,
September 12-15, 1994. Prepared by the
OECD Environment Directorate in
collaboration with the Directorate for
Science, Technology and Industry.
Includes bibliographical references.
Subjects: Food--Biotechnology--Safety
measures--Congresses. Transgenic
plants--Safety measures--Congresses.
Series: OECD documents LC
Classification: TP248.65.F66 F73 1996
Dewey Class No.: 363.19/2 20

Food testing reagents: impact of new
regulations and biotechnology / Philip
Rotheim. Published/Created: Norwalk,
CT: Business Communications Co.,
c1987. Related Authors: Rotheim,
Philip. Business Communications Co.

Description: xx, 217 leaves: ill.; 28 cm. ISBN: 0893367737 Notes: "June 1992"--T.p. verso. Subjects: Food testing reagents industry--United States. Biotechnology industries--United States. Food adulteration and inspection--Law and legislation United States. Market surveys--United states. Series: Business opportunity report; C-121 LC Classification: HD9000.9.U5 F598 1987

Forest and crop biotechnology: progress and prospects / Fredrick A. Valentine, editor. Published/Created: New York: Springer-Verlag, c1988. Related Authors: Valentine, Fredrick A. (Frederick Arthur), 1926- Description: xix, 466 p.: ill.; 25 cm. ISBN: 0387968482 (alk. paper) Notes: Papers from a colloquium held Apr. 18-20, 1985, at the State University of New York College of Environmental Science and Forestry, Syracuse, New York. Includes bibliographies and indexes. Subjects: Plant biotechnology--Congresses. Forestry biotechnology--Congresses. Trees--Biotechnology--Congresses. LC Classification: S494.5.B563 F67 1988 Dewey Class No.: 631 19

Forest products biotechnology / edited by Alan Bruce and John W. Palfreyman. Published/Created: London; Bristol, PA: Taylor & Francis, c1998. Related Authors: Bruce, A. M. Palfreyman, John W. Description: ix, 326 p.: ill.; 26 cm. ISBN: 0748404155 Notes: Includes bibliographical references and index. Subjects: Forest products--Biotechnology. Forestry biotechnology. Bioremediation. LC Classification: SD541 .F66 1998 Dewey Class No.: 634.9/8 21

Fostering the bioeconomic revolution in biobased products and bioenergy: an environmental approach / by the Biomass Research and Development Board. Published/Created: Golden, CO: Produced for the Board by the National Renewable Energy Laboratory, [2001] Related Authors: Biomass Research and Development Board (U.S.) National Renewable Energy Laboratory (U.S.) Description: 29 p.: col. ill., col. map; 28 cm. Notes: "January 2001." Cover title. Subjects: Biotechnology industries--United States. Biomass energy industries--United States. Pollution control industry--United States. Biotechnology--Research--United States. Biomass energy--Research--United States. Biological control systems--Research--United States. Biotechnology--Industrial applications--United States. Market surveys--United States. LC Classification: HD9999.B443 U638 2001 Dewey Class No.: 338.4/76606/0973 21

Genetic improvement of tomato / G. Kalloo, ed. Published/Created: Berlin; New York: Springer-Verlag, c1991. Related Authors: Kalloo, G. Description: xii, 358 p.: ill.; 25 cm. ISBN: 3540530622 (Berlin: acid-free paper) 0387530622 (New York: acid-free paper) Notes: Includes bibliographical references (p. [293]-352) and index. Subjects: Tomatoes--Breeding. Tomatoes--Genetics. Tomatoes--Biotechnology. Series: Monographs on theoretical and applied genetics; 14 LC Classification: SB349 .G42 1991 Dewey Class No.: 635/.64223 20

Genetic improvement of vegetable crops / edited by G. Kalloo, B.O. Bergh. Edition Information: 1st ed. Published/Created: Oxford; New York:

Pergamon Press, 1993. Related Authors: Kalloo, G. Bergh, B. O. Description: xi, 833 p.: ill.; 26 cm. ISBN: 0080408265: Notes: Includes bibliographical references and index. Subjects: Vegetables--Breeding. Vegetables--Genetics. Vegetables--Biotechnology. Crop improvement. LC Classification: SB324.7 .G46 1993 Dewey Class No.: 635/.0423 20

Genetic modification in the food industry: a strategy for food quality improvement / edited by Sibel Roller and Susan Harlander. Edition Information: 1st ed. Published/Created: London; New York: Blackie Academic & Professional, 1998. Related Authors: Roller, Sibel. Harlander, Susan K. Description: xvi, 260 p.: ill.; 24 cm. ISBN: 0751403997 Notes: Includes bibliographical references and index. Subjects: Food--Biotechnology. Genetically modified foods. LC Classification: TP248.65.F66 G455 1998 Dewey Class No.: 664 21

Genetically modified foods: safety issues / Karl-Heinz Engel, Gary R. Takeoka, Roy Teranishi, [editors]. Published/Created: Washington, DC: American Chemical Society, 1995. Related Authors: Engel, Karl-Heinz, 1954- Takeoka, Gary R. Teranishi, Roy, 1922- American Chemical Society. Division of Agricultural and Food Chemistry. American Chemical Society. Meeting (208th: 1994: Washington, D.C.) Description: x, 243 p.: ill.; 24 cm. ISBN: 0841233209 (acid-free, recycled paper) Notes: "Developed from a symposium sponsored by the Division of Agricultural and Food Chemistry at the 208th National Meeting of the American Chemical Society, Washington, DC, August 21-25, 1994." Includes bibliographical references and

indexes. Subjects: Genetically modified foods--Congresses. Food--Biotechnology--Congresses. Crops--Genetic engineering--Congresses. Series: ACS symposium series; 605 LC Classification: TP248.65.F66 G46 1995 Dewey Class No.: 363.19/2 20

Grogan, John. Genetically modified foods: 50 things you absolutely must know / John Grogan, with Sally Deneen. Published/Created: Emmaus, PA: Rodale, 2001. Projected Pub. Date: 0110 Related Authors: Deneen, Sally. Description: p. cm. ISBN: 0875968740 (alk. paper) Notes: Includes index. Subjects: Genetically modified foods--Popular works. Food--Biotechnology--Popular works. Genetically modified foods--Popular works. LC Classification: TP248.65.F66 G76 2001 Dewey Class No.: 664 21

Guenther, Kim. Biotechnology and sustainable agriculture: a bibliography / Kim Guenther; edited by Raymond Dobert and Jane Potter Gates. Published/Created: Beltsville, Md.: National Agricultural Library, [1994] Related Authors: Dobert, Raymond. Gates, Jane Potter. National Agricultural Library (U.S.) Description: iv, 32 p.: ill.; 28 cm. Notes: Shipping list no.: 94-0358-P. "September 1994." Includes indexes. Subjects: Sustainable agriculture--Bibliography. Agricultural biotechnology--Bibliography. Series: Special reference briefs; NAL-SRB. 94-13. Variant Series: Special reference briefs, 1052-536X; SRB 94-13 LC Classification: Z5074.S92 G84 1994 Dewey Class No.: 016.63 20

Haines, Russell. Biotechnology in forest tree improvement: with special reference to developing countries / by Russell Haines. Published/Created: Rome: Food and Agriculture Organization of the United Nations, 1994. Related Authors: Food and Agriculture Organization of the United Nations. Description: xxvii, 230 p.: ill.; 21 cm. ISBN: 9251034907 Notes: "M-31"--T.p. verso. Includes bibliographical references (p. 169-219). Subjects: Forestry biotechnology. Trees--Biotechnology. Trees--Breeding. Series: FAO forestry paper, 0258-6150; 118 LC Classification: SD387.B55 H35 1994 Dewey Class No.: 634.9/56 20

Halász, Anna. Use of yeast biomass in food production / authors, Anna Halász and Radomir Lásztity. Published/Created: Boca Raton: CRC Press, c1991. Related Authors: Lásztity, Radomír. Description: 312 p.: ill.; 27 cm. ISBN: 0849358663 Notes: Includes bibliographical references and index. Subjects: Single cell proteins--Biotechnology. Yeast fungi--Physiology. Food--Biotechnology. LC Classification: TP248.65.S56 H35 1990 Dewey Class No.: 664 20

Hill area development: issues and perspectives / editors, R.K. Rai ... [et al.]. Published/Created: Shillong: Geographical Society of North-Eastern Hill Region (India), 1990. Related Authors: Rai, Raj Kumar, 1940- Geographical Society of the North-Eastern Hill Region (India) Description: xvi, 228 p.: ill., maps; 23 cm. Summary: Contributed articles. Notes: Includes bibliographical references. Subjects: Environmental policy--India--Shillong. Rural development--India--Shillong. Agricultural biotechnology--India--Shillong. Shillong (India)--Economic policy--1981- LC Classification:

HC437.S624 H55 1990

Hobbelink, Henk. Biotechnology and the future of world agriculture / Henk Hobbelink. Published/Created: London; Atlantic Highlands, N.J.: Zed Books, c1989. Description: 159 p.: ill.; 23 cm. ISBN: 0862328365 0862328373 (pbk.) Notes: Includes bibliographical references and index. Subjects: Agricultural biotechnology--Forecasting. LC Classification: S494.5.B563 H63 1989 Dewey Class No.: 631 20

Hobbelink, Henk. New hope or false promise?: biotechnology and Third World agriculture / Henk Hobbelink. Edition Information: 2nd ed. Published/Created: Brussels, Belgium: International Coalition for Development Action, [1987] Related Authors: International Coalition for Development Action. Description: 72 p.: ill.; 21 cm. ISBN: 9072049012 Notes: "April 1987." Includes bibliographical references (p. 69-71). Subjects: Agricultural biotechnology. Agricultural biotechnology--Developing countries. Agriculture--Developing countries. Biotechnology industries. LC Classification: S494.5.B563 H64 1987

Horticultural Biotechnology Symposium (1989: University of California, Davis) Horticultural biotechnology: proceedings of the Horticultural Biotechnology Symposium, held at the University of California, Davis, California, August 21-23, 1989 / editors, Alan B. Bennett, Sharman D. O'Neill. Published/Created: New York: Wiley-Liss, c1990. Related Authors: Bennett, Alan B. O'Neill, Sharman D. Description: xvii, 387 p.: ill.; 24 cm. ISBN: 0471568066 Notes: Includes bibliographical references and index. Subjects: Plant biotechnology--Congresses. Horticulture--Congresses. Series: Plant biology; v. 11 LC Classification: S494.5.B563 H67 1989 Dewey Class No.: 631.5/2 20

International initiatives in biotechnology for developing country agriculture: promises and problems. Published/Created: Paris: Organisation for Economic Co-operation and Development, 1994. Related Authors: Organisation for Economic Co-operation and Development. Description: 60 p.: ill.; 30 cm. Notes: Includes bibliographical references (p. 55-56). Subjects: Agricultural biotechnology--Developing countries. Series: OECD working papers; v. 2, no. 53. Variant Series: OECD working papers, 1022-2227; vol. 2, no. 53 LC Classification: HD72 .O38 vol. 2, no. 53

International Symposium on Applications of Biotechnology to Tree Culture, Protection and Utilization (2nd: 1994: Bloomington, Minn.) Papers and abstracts presented at the 1994 Second Interntional Symposium on the Applications of Biotechnology to Tree Culture, Protection, and Utilization: Bloomington, Minnesota, October 2-6, 1994 / edited by Charles H. Michler ... [et al.]; sponsored by USDA Office of Agricultural Biotechnology ... [et al.]; hosted by USDA Forest Service, North Central Forest Experiment Station. Published/Created: St. Paul, Minn.: The Station, 1994. Related Authors: Michler, Charles H. United States. Office of Agricultural Biotechnology. North Central Forest Experiment Station (Saint Paul, Minn.) Related Titles: Applications of biotechnology to tree culture, protection, and utilization.

Description: 203 p.: ill.; 28 cm. Notes: Includes bibliographical references. Subjects: Forestry biotechnology--Congresses. Series: General technical report NC; 175 LC Classification: SD387.B55 I57 1994

International workshop on biotechnology in agriculture: evolving a research agenda for the ICGEB, September, 17-22, 1985 / editors, S. Natesh, V.L. Chopra, S. Ramachandran. Published/Created: New Delhi: Oxford & IBH Pub. Co., c1987. Related Authors: Natesh, S. Chopra, V. L. Ramachandran, S., 1934- United Nations Industrial Development Organization. India. Dept. of Science and Technology. International Workshop on Biotechnology in Agriculture: Evolving a Research Agenda for the ICGEB (1985: New Delhi, India) Description: 321 p.: ill.; 25 cm. ISBN: 8120402413: Summary: On plant molecular biology and biotechnology. Notes: Papers presented at the International Workshop on Biotechnology in Agriculture: Evolving a Research Agenda for the ICGEB, New Delhi, 1985; organized by United Nations Industrial Development Organization and Dept. of Science and Technology, India. Includes bibliographies. Subjects: Plant biotechnology--Congresses. Agricultural biotechnology--Congresses. LC Classification: S494.5.B563 I58 1985 Dewey Class No.: 631 19

Juma, Calestous. The gene hunters: biotechnology and the scramble for seeds / Calestous Juma. Published/Created: Princeton, N.J.: Princeton University Press, 1989. Description: xiv, 288 p.: maps; 23 cm. ISBN: 0691042586: 0691003785 (pbk.)

Notes: Includes index. Bibliography: p. 249-272. Subjects: Plant biotechnology--Africa--Forecasting. Germplasm resources, Plant--Africa--Utilization. Developing countries--Economic policy. Series: African Centre for Technology Studies research series; no. 1 LC Classification: S494.5.B563 J86 1989 Dewey Class No.: 338.1/6 19

Lee, B. H. (Byong H.) Fundamentals of food biotechnology / Byong H. Lee. Published/Created: New York: VCH, c1996. Description: xvi, 431 p.: ill.; 24 cm. ISBN: 1560816945 (cloth: acid-free paper) Notes: Includes bibliographical references and index. Subjects: Food--Biotechnology. Series: Food science and technology (VCH Publishers) Variant Series: Food science and technology LC Classification: TP248.65.F66 L44 1996 Dewey Class No.: 664/.024 20

Legumes and oilseed crops I / edited by Y.P.S. Bajaj. Published/Created: Berlin; New York: Springer-Verlag, c1990. Related Authors: Bajaj, Y. P. S., 1936- Related Titles: Legumes and oilseed crops one. Description: xx, 682 p.: ill.; 25 cm. ISBN: 0387507868 (U.S.: alk. paper) Notes: Includes bibliographical references and index. Subjects: Legumes--Micropropagation. Oilseed plants--Micropropagation. Legumes--Biotechnology. Oilseed plants--Biotechnology. Series: Biotechnology in agriculture and forestry; 10 LC Classification: SB177.L45 L45 1990 Dewey Class No.: 633.3/042 20

Maize / edited by Y.P.S. Bajaj. Published/Created: Berlin; New York: Springer-Verlag, c1994. Related Authors: Bajaj, Y. P. S., 1936-

Description: xxvi, 632 p.: ill.; 25 cm. ISBN: 354056392X (Berlin: acid-free paper): 038756392X (New York: acid-free paper) Notes: Includes bibliographical references and index. Subjects: Corn--Micropropagation. Corn--Biotechnology. Series: Biotechnology in agriculture and forestry; 25 LC Classification: SB191.M2 M3247 1994 Dewey Class No.: 631.5/23 20

Managing agricultural biotechnology: addressing research program needs and policy implications / edited by Joel I. Cohen. Published/Created: Wallingford, Oxon, UK; New York: CABI Pub. in association with the International Service for National Agricultural Research, c1999. Related Authors: Cohen, Joel I., 1950- Description: xxi, 323 p.: ill.; cm. ISBN: 0851994008 Notes: Includes bibliographical references. Subjects: Agricultural biotechnology--Management. Series: Biotechnology in agriculture series; 23 LC Classification: S494.5.B563 M36 1999 Dewey Class No.: 631.5/233 21

Marshall, Elizabeth L. High-tech harvest: a look at genetically engineered foods / by Elizabeth L. Marshall. Published/Created: New York: Franklin Watts, c1999. Description: 144 p.: ill.; 24 cm. ISBN: 0531114341 Summary: An overview of recombined DNA technology, or genetic engineering, techniques used to create crop plants and farm animals with characteristics that are attractive to farmers, food processors, and consumers. Notes: Includes bibliographical references (p. 128-140) and index. Subjects: Agricultural biotechnology--Juvenile literature. Genetically modified foods--

Juvenile literature. Food--Biotechnology--Juvenile literature. Agricultural biotechnology. Genetic engineering. Food--Biotechnology. Variant Series: An Impact book LC Classification: S494.5.B563 M368 1999 Dewey Class No.: 641.3 21

März, U. Biotechnology production of food and feed ingredients: a global view / Ulrich Marz, project analyst. Published/Created: Norwalk, CT: Business Communications Co., c1997. Related Authors: Business Communications Co. Description: xxi, 138 leaves: ill.; 28 cm. ISBN: 1569650217 Subjects: Biotechnology industries. Biotechnology--Industrial applications. Food industry and trade. Feed industry. Market surveys. Series: Business opportunity report; B-109 LC Classification: HD9999.B442 M37 1997

Mather, Robin. A garden of unearthly delights: bioengineering and the future of food / Robin Mather. Published/Created: New York, N.Y., U.S.A.: Penguin Books, c1995. Description: xiii, 205 p.; 23 cm. ISBN: 0525938648 (acid-free): Notes: "A Dutton book." Includes bibliographical references (p. [191]-195) and index. Subjects: Agricultural biotechnology. Food--Biotechnology. Food. Nutrition. LC Classification: S494.5.B563 M38 1995 Dewey Class No.: 338.1/9 20

McHughen, Alan. Pandora's picnic basket: the potential and hazards of genetically modified foods / Alan McHughen. Published/Created: Oxford; New York: Oxford University Press, 2000. Description: viii, 277 p.: ill.; 25 cm. ISBN: 0198507143 (Pbk.) 0198506740 (hbk.) Notes: Includes bibliographical references and index. Subjects:

Genetically modified foods. Food--Biotechnology. Transgenic plants. LC Classification: TP248.65.F66 M37 2000 Dewey Class No.: 363.19/29 21

NATO Advanced Research Workshop on Woody Plant Biotechnology (1989: Placerville, Calif.) Woody plant biotechnology / edited by M.R. Ahuja. Published/Created: New York: Plenum Press, c1991. Related Authors: Ahuja, M. R., 1933- North Atlantic Treaty Organization. Scientific Affairs Division. Description: xi, 373 p.: ill.; 26 cm. ISBN: 0306440199 Notes: "Published in cooperation with NATO Scientific Affairs Division." "Proceedings of a NATO Advanced Research Workshop on Woody Plant Biotechnology, held October 15-19, 1989, in Placerville, California"--T.p. verso. Includes bibliographical references and index. Subjects: Forestry biotechnology--Congresses. Trees--Biotechnology--Congresses. Woody plants--Biotechnology--Congresses. Series: NATO ASI series. Series A, Life sciences; v. 210 LC Classification: SD387.B55 N37 1989 Dewey Class No.: 634.9 20

Nayudamma Memorial Symposium (1986: Madras, India) Agricultural applications of biotechnology: proceedings of the Nayudamma Memorial Symposium, 15-17 December 1986 / edited by A.N. Rao, H.Y. Mohan Ram. Published/Created: Madras, India: COSTED, 1987. Related Authors: Nayudamma, Y. Rao, A. N. Mohan Ram, H. Y. International Council of Scientific Unions. Committee on Science and Technology in Developing Countries. Description: 210 p.: ill., port.; 26 cm. ISBN: 9971621568: Notes: Organized by the Committee on

Science and Technology in Developing Countries, et al. Includes bibliographical references. Subjects: Agricultural biotechnology--Congresses. Agricultural biotechnology--Developing countries Congresses. LC Classification: S494.5.B563 N39 1987 Dewey Class No.: 630 20

Perspectives on agricultural biotechnology: proceedings from "Biotechnology: Progress or Problem?": a conference for developing community leaders, Binghamton, New York, January 17-19, 2001. Published/Created: Ithaca, N.Y: Natural Resource, Agriculture, and Engineering Service, 2001. Projected Pub. Date: 0205 Related Authors: Natural Resource, Agriculture, and Engineering Service. Cooperative Extension. Description: p. cm. ISBN: 0935817700 (pbk.) Subjects: Biotechnology--Congresses. Biotechnology--Social aspects--Congresses. Agricultural biotechnology--Congresses. Series: NRAES (Series); 144 Variant Series: NRAES; 144 LC Classification: TP248.14 .B55927 2002 Dewey Class No.: 303.48/3 21

Plant biochemical regulators / edited by Harold W. Gausman. Published/Created: New York: M. Dekker, c1991. Related Authors: Gausman, H. W. Description: viii, 363 p.: ill.; 24 cm. ISBN: 0824785363 (acid-free paper) Notes: Includes bibliographical references and index. Subjects: Plant regulators. Plant biotechnology. Series: Books in soils, plants, and the environment LC Classification: SB128 .P48 1991 Dewey Class No.: 631.8 20

Plant biotechnologies for developing countries: proceedings of an

international symposium organized jointly by the Technical Centre for Agricultural and Rural Co-operation (CTA) and the Food and Agriculture Organization of the United Nations (FAO) and held in Luxembourg, 26-30 June 1989 / edited by A. Sasson and V. Costarini. Published/Created: [Ede, Netherlands]: CTA; [Rome, Italy]: FAO, [1989] Related Authors: Sasson, Albert. Costarini, Vivien. Technical Centre for Agricultural and Rural Cooperation (Ede, Netherlands) Food and Agricultural Organization of the United Nations. Description: xiv, 368 p.; 25 cm. ISBN: 9290810750 Notes: Includes bibliographical references and index. Subjects: Plant biotechnology--Congresses. Plant biotechnology--Developing countries--Congresses. LC Classification: SB106.B56 P58 1989 Dewey Class No.: 631.5/23 20

Plant biotechnology. Published/Created: Glasgow: Blackie; New York: Chapman and Hall, c1991-c1993. Description: 3 v.: ill.; 24 cm. Vol. 1-v. 3. Current Frequency: Irregular Cancel/Invalid LCCN: sn 92015072 CODEN: PBSEE8 Indexed entirely by: Bibliography of agriculture 0006-1530 Indx'd selectively by: Chemical abstracts 0009-2258 Subjects: Plant biotechnology. LC Classification: CLASSED SEPARATELY

Pollen biotechnology for crop production and improvement / edited by K.R. Shivanna, V.K. Sawhney. Published/Created: Cambridge [England]; New York: Cambridge University Press, 1997. Related Authors: Shivanna, K. R. Sawhney, V. K. Description: xiv, 448 p.: ill.; 24 cm. ISBN: 052147180X Notes: Includes bibliographical references and index.

Subjects: Pollen--Biotechnology. Plant breeding. Crops--Genetic engineering. Crop improvement. LC Classification: SB106.B56 P65 1997 Dewey Class No.: 631.5/3 20

Potato / edited by Y.P.S. Bajaj. Published/Created: Berlin; New York: Springer-Verlag, c1987. Related Authors: Bajaj, Y. P. S., 1936- Description: xxii, 509 p.: ill.; 25 cm. ISBN: 3540179666 (Berlin) 0387179666 (New York) Notes: Includes bibliographical references and index. Subjects: Potatoes-- Biotechnology. Series: Biotechnology in agriculture and forestry; 3 LC Classification: SB211.P8 P77 1987 Dewey Class No.: 633/.491 19

Public hearing on should New York State regulate genetically modified organisms (GMO's) and foods produced from these GMO's / Assembly Standing Committee on Agriculture ... [et al.]; reported by Gabrielle Paige Hackett. Published/Created: [Ithaca?]: Paige Reporting, [2000] Related Authors: Hackett, Gabrielle Paige. New York (State). Legislature. Assembly. Standing Committee on Agriculture. Description: 207 leaves; 29 cm. Notes: At head of State of New York, New York State Assembly. Hearing was held on Oct. 3, 2000 in Ithaca, N.Y. Assemblyman William Magee, chairman, Assembly Agriculture Committee. Subjects: Genetic engineering--Government policy--New York (State) Transgenic organisms--Government policy--New York (State) Genetically modified foods--Government policy--New York (State) Agricultural biotechnology-- Government policy--New York (State) LC Classification: KFN5010.4 .A395 2000b

Public hearing on should New York State regulate genetically modified organisms (GMO's) and foods produced from these GMO's / Assembly Standing Committee on Agriculture ... [et al.]; reported by Melissa L. Nesbitt. Published/Created: [Ithaca?]: Paige Reporting, [2000] Related Authors: Nesbitt, Melissa L. New York (State). Legislature. Assembly. Standing Committee on Agriculture. Description: 179 leaves; 29 cm. Notes: At head of State of New York, New York State Assembly. Hearing was held on Oct. 5, 2000 in Rochester, N.Y. Assemblyman William Magee, chairman, Assembly Agriculture Commitee. Subjects: Transgenic organisms--Government policy--New York (State) Genetically modified foods--Government policy--

Qaim, Matin. Potential impacts of crop biotechnology in developing countries / Matin Qaim. Published/Created: Frankfurt am Main; New York: Peter Lang, c2000. Description: xiv, 168 p.: ill., maps; 22 cm. ISBN: 0820448303 (pbk.) Notes: Includes bibliographical references (p. [155]-168). Subjects: Agricultural biotechnology--Developing countries. Agricultural innovations-- Developing countries. Agricultural biotechnology--Economic aspects-- Developing countries. Agricultural innovations--Economic aspects-- Developing countries. Series: Development economics and policy; Bd. 17 Variant Series: Development economics and policy, 0948-1338; vol. 17 LC Classification: S494.5.B563 Q35 2000 Dewey Class No.: 338.1/6/091724 21

Rechcigl, Jack E. Biological and biotechnological control of insect pests / Jack E. Rechcigl, Nancy A. Rechcigl. Published/Created: Boca Raton, Fla.: Lewis Publishers, c1998. Related Authors: Rechcigl, Nancy A. Description: 374 p.: ill.; 25 cm. ISBN: 1566704790 (alk. paper) Notes: Includes bibliographical references. Subjects: Insect pests--Biological control. Biological pest control agents. Agricultural biotechnology. Series: Agriculture & environment series. Variant Series: Agriculture and environment series LC Classification: SB933.3 .R436 1998 Dewey Class No.: 632/.96 21

Regional Workshop on Biotechnology in Agriculture (1986: Los Baños, Philippines) Proceedings / Regional Workshop on Biotechnology in Agriculture, October 8-11, 1986, SEARCA, Los Baños, Laguna, Philippines. Published/Created: [Los Baños, Philippines]: SEAMEO Regional Center for Graduate Study and Research in Agriculture, [1986] Related Authors: SEAMEO Regional Center for Graduate Study and Research in Agriculture. Description: vi, 317 p.: ill.; 26 cm. Notes: Includes bibliographical references. Subjects: Agricultural biotechnology--Asia, Southeastern Congresses. Agricultural biotechnology--Congresses. LC Classification: S494.5.B563 R44 1986

Safety considerations for biotechnology: scale-up of crop plants. Published/Created: Paris, France: Organisation for Economic Co-operation and Development; Washington, D.C.: OECD Publications and Information Centre [distributor], 1993. Related Authors: Organisation for

Economic Co-operation and Development. Description: 40 p.; 23 cm. ISBN: 9264140441 Notes: Published in French under the Considérations de sécurité relatives a la biotechnologie: passage a l'échelle supérieure des plantes cultivées. Includes bibliographical references (p. 35-36). Subjects: Plant biotechnology--Safety measures. Genetic engineering--Risk assessment. LC Classification: SB106.B56 S24 1993

Safety considerations for biotechnology: scale-up of micro-organisms as biofertilizers. Published/Created: Paris: Organisation for Economic Co-operation and Development; Washington, D.C.: OECD Publications and Information Centre [distributor], c1995. Related Authors: Organisation for Economic Co-operation and Development. Directorate for Science, Technology and Industry. Organisation for Economic Co-operation and Development. Environment Directorate. Description: 67 p.; 23 cm. ISBN: 9264143440 Notes: Prepared by the OECD Directorate for Science, Technology and Industry in collaboration with Environment Directorate. Includes bibliographical references (p. 53-59). Subjects: Biofertilizers. Microbial biotechnology--Safety measures. LC Classification: S654.5 .S24 1995

Safety evaluation of foods derived by modern biotechnology: concepts and principles. Published/Created: Paris: Organisation for Economic Co-operation and Development, c1993. Related Authors: Organisation for Economic Co-operation and Development. Environment Directorate. Organisation for Economic Co-operation and Development. Directorate for Science, Technology, and Industry. Organisation for Economic Co-operation and Development. Group of National Experts on Safety and Biotechnology. Description: 79 p.; 23 cm. ISBN: 9264138595 (pbk.): Notes: "Prepared by the OECD Environment Directorate, in collaboration with the Directorate for Science, Technology and Industry. It is the product of work undertaken by the Group of National Experts on Safety in Biotechnology"--P. 3. Includes bibliographical references (p. 71-72). Subjects: Food--Biotechnology--Safety measures. LC Classification: TP248.65.F66 S24 1993 Dewey Class No.: 363.19/29 20

Schor, Joel. The evolution and development of biotechnology: a revolutionary force in American agriculture / Joel Schur. Published/Created: Washington, DC: U.S. Dept. of Agriculture, Economic Research Service, Agriculture and Rural Economy Division, [1994] Related Authors: United States. Dept. of Agriculture. Economic Research Service. Agriculture and Rural Economy Division. Description: v, 134 p.; 28 cm. Notes: Cover title. Includes bibliographical references. Subjects: Agricultural biotechnology--Research--United States. Series: ERS staff report; AGES 9424. Variant Series: Staff report; AGES 9424 LC Classification: S494.5.B563 S34 1994

Schulz, Evan, 1967- The opportunities and hazards of agricultural biotechnology / Evan Schulz. Published/Created: Washington, DC: Economic Strategy Institute, 2000. Description: 45 p.; 23 cm. ISBN: 1888773049 Notes: Includes bibliographical references. Subjects: Agricultural biotechnology. LC

Classification: S494.5.B563 S37 2000 Dewey Class No.: 338.1/6 21

Science Council of Canada. Seeds of renewal: biotechnology and Canada's resource industries / [Science Council of Canada]. Published/Created: Ottawa, Ont.: The Council, c1985. Description: 94 p.: ill.; 25 cm. ISBN: 0660119145 (pbk.): Notes: Bibliography: p. 76-82. Subjects: Biotechnology--Canada. Natural resources--Canada. Series: Report (Science Council of Canada); 38. Variant Series: Report; 38 LC Classification: TP248.2 .S35 1985 Dewey Class No.: 338.4/76606 19

Science for the food industry of the 21st century: biotechnology, supercritical fluids, membranes and other advanced technologies for low calorie, healthy food alternatives / edited by Manssur Yalpani. Published/Created: Mount Prospect, Il.: ATL Press, c1993. Related Authors: Yalpani, Mansur, 1950- Related Titles: Biotechnology, supercritical fluids, membranes and other advanced technologies for low calorie, healthy food alternatives. Description: 414 p.: ill.; 24 cm. ISBN: 1882360451 Contents: Low calorie, healthy foods -- New food ingredients: High potency sweetener properties; Fat macronutrient substitutes; Sugar macronutrient substitutes -- Advanced technologies: Biotechnological approaches; Separation technologies; Stabilization technologies -- Metabolic, safety and regulatory issues: Metabolism; safety; Regulatory issues. Notes: Includes bibliographical references and index. Subjects: Food adulteration and inspection. Food--Analysis. Biotechnology. Food industry and trade. Series: Frontiers in foods and food ingredients, 1072-429X; vol. 1 LC

Classification: TX531 .S38 1993 Dewey
Class No.: 363.19/2 20

Spangenberg, G. (German), 1959-
Biotechnology in forage and turf grass
improvement / G. Spangenberg, Z.-Y.
Wang, I. Potrykus. Published/Created:
Berlin; New York: Springer, c1998.
Related Authors: Wang, Z.-Y. (Zeng-
Yu), 1963- Potrykus, I. (Ingo), 1933-
Description: ix, 200 p.: ill. (some col.);
25 cm. ISBN: 3540638261 (hardcover)
Notes: Includes bibliographical
references and index. Subjects: Forage
plants--Biotechnology. Turfgrasses--
Biotechnology. Series: Monographs on
theoretical and applied genetics; 23 LC
Classification: SB193 .S668 1998
Dewey Class No.: 633.2 21

Strategies for assessing the safety of foods
produced by biotechnology: report of a
joint FAO/WHO consultation.
Published/Created: Geneva: World
Health Organization, 1991. Related
Authors: Food and Agriculture
Organization of the United Nations.
World Health Organization.
Description: iv, 59 p.; 24 cm. ISBN:
9241561459 Notes: "Joint FAO/WHO
Consultation on the Assessment of
Biotechnology in Food Production and
Processing as Related to Food Safety ...
held in Geneva from 5 to 10 November
1990"--Introd. Includes bibliographical
references (p. 53-54). Subjects: Food--
Biotechnology. Food--Toxicology. LC
Classification: TP248.65.F66 S77 1991
Dewey Class No.: 363.19/2 20

The Impact of biotechnology on agriculture
in the European Community to the year
2005: study prepared for the Directorate
General for Agriculture / Commission
of the European Communities.
Published/Created: Luxembourg: Office

for Official Publications of the
European Communities; Washington,
DC: European Community Information
Service [distributor], 1989. Related
Authors: Commission of the European
Communities. Directorate-General for
Agriculture. Description: ii, 161 p.: ill.;
30 cm. ISBN: 9282588491 Subjects:
Agricultural biotechnology--European
Economic Community countries--
Forecasting. LC Classification:
S494.5.B563 I48 1989 Dewey Class
No.: 338.1/6/094 20

The Impact of biotechnology on pesticides.
Published/Created: New York, N.Y.
(106 Fulton St., New York 10038):
Frost & Sullivan, c1989. Related
Authors: Frost & Sullivan. Related
Titles: Biotechnology-related alternative
pesticides. Description: viii, 350 p.: ill.;
29 cm. Notes: Spine Biotechnology-
related alternative pesticides. "Winter
1989." "A2150." Subjects: Pesticides
industry--Technological innovations--
United States. Biotechnology--United
States--Industrial applications. Market
surveys--United States. LC
Classification: HD9660.P33 U535 1989

The Maize handbook / Michael Freeling,
Virginia Walbot, editors.
Published/Created: New York:
Springer-Verlag, c1994. Related
Authors: Freeling, Michael. Walbot,
Virginia. Description: xxvi, 759 p.: ill.
(some col.); 25 cm. ISBN: 0387978267
(New York) 3540978267 (Berlin)
Notes: Includes bibliographical
references and index. Subjects: Corn--
Handbooks, manuals, etc. Corn--
Biotechnology--Handbooks, manuals,
etc. LC Classification: SB191.M2
M3275 1994 Dewey Class No.: 633.1/5
20

Toyota Conference (12th: 1998: Shizuoka-shi, Japan) Proceedings of the 12th Toyota Conference: challenge of plant and agricultural sciences to the crisis of biosphere on the earth in the 21st century / Kazuo Watanabe and Atsushi Komamine. Published/Created: Georgetown, TX: R.G. Landes Co., c2000. Related Authors: Watanabe, Kazuo N. Komamine, Atsushi, 1929- Description: 309 p.: ill. (some col.); 24 cm. ISBN: 1570596166 (alk. paper) Notes: Includes bibliographical references and index. Subjects: Crops--Genetic engineering--Congresses. Plant biotechnology--Congresses. Series: Environmental intelligence unit LC Classification: SB123.57 .T69 1998 Dewey Class No.: 631.5/233 21

Trends in food science & technology. Published/Created: Cambridge, UK: Elsevier Trends Journals, c1990- Description: v.: ill.; 30 cm. Vol. 1, no. 1 (July 1990)- Current Frequency: Monthly ISSN: 0924-2244 CODEN: TFTEEH Notes: Title from cover. SERBIB/SERLOC merged record Compilation of articles reprinted in the reference ed. Indexed entirely by: Bibliography of agriculture 0006-1530 Vol. 3, No. 12, 1992 Indx'd selectively by: Chemical abstracts 0009-2258 1990- Subjects: Food industry and trade--Periodicals. Food--Biotechnology--Periodicals. Food Industry--trends--Periodicals. Food Technology--trends--Periodicals. LC Classification: TP368 .T74 Dewey Class No.: 664 20

Tropical tuber crops: problems, prospects and future strategies / editors, G.T. Kurup ... [et al.]. Published/Created: New Delhi: Oxford & IBH Pub. Co., c1996. Related Authors: Kurup, G. T. Description: xiii, 597 p.: ill.; 24 cm.

ISBN: 8120410327 Summary: Contributed articles. Notes: Includes bibliographical references.

United States. Congress. Joint Economic Committee. New directions for agriculture: the science and technology of the future: hearing before the Joint Economic Committee, Congress of the United States, Ninety-eighth Congress, second session, October 2, 1984. Published/Created: Washington: U.S. G.P.O., 1984. Description: iii, 147 p.: ill.; 24 cm. Notes: Distributed to some depository libraries in microfiche. Item 1000-B, 1000-C (microfiche) Includes bibliographies. Subjects: Agricultural innovations--United States. Plant introduction--United States. Agricultural biotechnology--United States. Series: United States. Congress. Senate. S. hrg.; 98-1129. Variant Series: S. hrg.; 98-1129 LC Classification: KF25 .E2 1984k Dewey Class No.: 338.1/0973 19

United States. Congress. Senate. Committee on Agriculture, Nutrition, and Forestry. Agricultural consolidation antitrust: hearing before the Committee on Agriculture, Nutrition, and Forestry, United States Senate, OneHundred Sixth Congress, first session ... July 27, 1999. Published/Created: Washington: U.S. G.P.O.: For sale by the U.S. G.P.O., Supt. of Docs., Cogressional Sales Office, 1999. Description: iii, 85 p.: ill.; 24 cm. ISBN: 0160595665 Notes: Distributed to some depository libraries in microfiche. Shipping list no.: 2000-0025-P. Includes bibliographical references. Subjects: Agricultural laws and legislation--United States. Antitrust law--United States. Agricultural industries--Mergers--United States. Biotechnology industries--Mergers--

United States. Series: United States.
Congress. Senate. S. hrg.; 106-126.
Variant Series: S. hrg.; 106-126 LC
Classification: KF26 .A35 1999a
Dewey Class No.: 338.8/263/0973 21

United States. Congress. Senate. Committee
on Agriculture, Nutrition, and Forestry.
Agricultural research and development:
hearings before the Committee on
Agriculture, Nutrition, and Forestry,
United States Senate, One Hundred
Sixth Congress, first session ... October
6, 7, 1999. Published/Created:
Washington: U.S. G.P.O.: For sale by
the U.S. G.P.O., Supt. of Docs.,
Congressional Sales Office, 2000.
Description: v, 306 p.: ill.; 24 cm.
ISBN: 0160605059 Notes: "Printed for
the use of the Committee on
Agriculture, Nutrition, and Forestry."
Distributed to some depository libraries
in microfiche. Shipping list no.: 2000-
0245-P. Includes bibliographical
references. Subjects: Agricultural
biotechnology--Research--United
States. Series: United States. Congress.
Senate. S. hrg.; 106-455. Variant Series:
S. hrg.; 106-455 LC Classification:
KF26 .A35 1999p

United States. Congress. Senate. Committee
on Agriculture, Nutrition, and Forestry.
Alternative Agricultural Products
Research Act of 1987: hearing before
the Committee on Agriculture,
Nutrition, and Forestry, United States
Senate, One hundredth Congress, first
session on S. 970, a bill to authorize a
research program for the modification
of plants focusing on the development
and production of new marketable
industrial and commercial products, and
for other purposes, May 15, 1987.
Published/Created: Washington: U.S.
G.P.O.: For sale by the Supt. of Docs.,

Congressional Sales Office, U.S.
G.P.O., 1987. Description: iv, 156 p.:
ill.; 24 cm. Notes: Distributed to some
depository libraries in microfiche.
Shipping list no.: 87-600-P. Item 1032-
C, 1032-D (microfiche) Subjects: Plant
biotechnology--Research--Law and
legislation--United States. Series:
United States. Congress. Senate. S. hrg.;
100-208. Variant Series: S. hrg.; 100-
208 LC Classification: KF26 .A35
1987c Dewey Class No.: 343.73/076999
347.30376999 19

United States. Congress. Senate. Committee
on Agriculture, Nutrition, and Forestry.
Trade issues: hearing before the
Committee on Agriculture, Nutrition,
and Forestry, United States Senate, One
Hundred Sixth Congress, first session,
on trade issues, June 24, 1999.
Published/Created: Washington: U.S.
G.P.O.: For sale by the U.S. G.P.O.,
Supt. of Docs., Congressional Sales
Office, 2000. Description: iii, 87 p.; 24
cm. ISBN: 016060172X Notes:
Distributed to some depository libraries
in microfiche. Shipping list no.: 2000-
0166-P. Subjects: Agriculture--United
States. Produce trade--United States.
Agricultural biotechnology--United
States. International trade. Exports--
United States. United States--
Commercial treaties. Series: United
States. Congress. Senate. S. hrg.; 106-
334. Variant Series: S. hrg.; 106-334 LC
Classification: KF26 .A35 1999i

United States. Congress. Senate. Committee
on Agriculture, Nutrition, and Forestry.
The use of regulation of biotechnology
in agriculture: joint hearing before the
Committee on Agriculture, Nutrition,
and Forestry and the Subcommittee on
Technology and the Law of the
Committee on the Judiciary, United

States Senate, One Hundredth Congress, first session, on the potential of biotechnology and America's competitive position, November 4, 1987. Published/Created: Washington: U.S. G.P.O.: For sale by the Supt. of Docs., Congressional Sales Office, U.S. G.P.O., 1988. Related Authors: United States. Congress. Senate. Committee on the Judiciary. Subcommittee on Technology and the Law. Description: iv, 113 p.: ill.; 24 cm. Notes: Distributed to some depository libraries in microfiche. Shipping list no.: 88-304-P. Item 1032-C, 1032-D (microfiche) Subjects: Agricultural biotechnology--United States. Agricultural innovations--United States. Genetic engineering--United States. Series: United States. Congress. Senate. S. hrg.; 100-565. Variant Series: S. hrg.; 100-565 LC Classification: KF26 .A35 1987h Dewey Class No.: 630/.2/574 19

United States. Congress. Senate. Committee on Agriculture, Nutrition, and Forestry. Subcommittee on Agricultural Research and General Legislation. Implementation of the Alternative Agricultural Research and Commercialization (AARC) Act of 1990: hearing before the Subcommittee on Agricultural Research and General Legislation of the Committee on Agriculture, Nutrition, and Forestry, United States Senate, One Hundred Second Congress, second session ... focusing on the current activities of the AARC Board and future activities, September 29, 1992. Published/Created: Washington: U.S. G.P.O.: For sale by the U.S. G.P.O., Supt. of Docs., Congressional Sales Office, 1993. Description: iii, 74 p.; 24 cm. ISBN: 0160399300 Notes: Distributed to some depository libraries

in microfiche. Shipping list no.: 93-0173-P. Includes bibliographical references (p. 50). Subjects: Agricultural biotechnology--Research--Government policy United States. Agricultural innovations--Research--Government policy United States. Plant products--Research--Government policy--United States. Agriculture--Research--Government policy--United States. New products--Research--Government policy--United States. Series: United States. Congress. Senate. S. hrg.; 102-946. Variant Series: S. hrg.; 102-946 LC Classification: KF26 .A3534 1992d

United States. General Accounting Office. International trade: concerns over biotechnology challenge U.S. agricultural exports: report to the Ranking Minority Member, Committee on Finance, U.S. Senate / United States General Accounting Office. Published/Created: Washington, D.C. (P.0. Box 37050 Washington 20013): The Office, [2001] Related Authors: United States. Senate. Committee on Finance. Description: 29 p.; 28 cm. Notes: Cover title. "June 2001." "GAO-01-727." Includes bibliographical references. Subjects: Biotechnology industries--United States. Agricultural biotechnology--United States. Agricultural biotechnology--Public opinion. Exports--United States. Competition, International. LC Classification: HD9999.B443 U677 2001

Warmbrodt, Robert D. World list of serials in agricultural biotechnology / by Robert D. Warmbrodt and Diana Airozo; edited by Stanislaw J. Kosecki and David H. Goldberg. Published/Created: Beltsville, Md.: U.S.

Dept. of Agriculture, National Agricultural Library, 1993. Related Authors: Airozo, Diana. Kosecki, Stanislaw J. Goldberg, David H. National Agricultural Library (U.S.) Description: ix, 471 p.; 28 cm. Notes: Shipping list no.: 93-0579-P. "July 1993." Includes indexes. Subjects: Agricultural biotechnology--Periodicals--Bibliography. Series: Bibliographies and literature of agriculture; no. 116 LC Classification: Z5071.A1 W37 1993

Weck, Edward. Transgenic plants: production and commodity solutions / author, Edward Weck; managing editor, Susan C. DiClemente. Published/Created: Southborough, MA: International Business Communications, c1998. Related Authors: DiClemente, Susan C. Description: 1 v. (in various pagings): ill.; 28 cm. ISBN: 1579361269 Notes: Includes bibliographical references. Subjects: Genetic engineering industry--United States. Biotechnology industries--United States. Agricultural industries--United States. Food industry and trade--United States. Primary commodities--United States. Genetically modified foods--United States. Transgenic plants. Market surveys--United States. Series: D & MD reports; #1985. Variant Series: D&MD reports; #1985 LC Classification: HD9999.G453 U69 1998

Wheat, David. Biotechnology in agriculture: impacts on products and markets / David Wheat. Published/Created: Waltham, MA: Decision Resources, [c1995] Related Authors: Decision Resources, Inc. Description: ix, 121 leaves: ill.; 30 cm. Subjects: Biotechnology industries. Agricultural biotechnology. Market surveys. Series:

DR reports LC Classification: HD9999.B442 W47 1995

Wiggert, Lara. Biotechnology, food science and technology: January 1991 - March 1993 / Lara Wiggert and Robert Warmbrodt. Published/Created: Beltsville, Md.: National Agricultural Library, [1993] Related Authors: Warmbrodt, Robert D.

BIOLOGY

Actinomycetes in biotechnology / edited by M. Goodfellow, S.T. Williams, M. Mordarski. Published/Created: San Diego: Academic Press, 1988. Related Authors: Goodfellow, M. Williams, S. T. (Stanley Thomas), 1937- Mordarski, Marian. Description: 501 p.: ill.; 24 cm. ISBN: 0122896734 Notes: Includes bibliographical references. Subjects: Actinomycetales--Biotechnology. LC Classification: QR82.A35 A28 1988 Dewey Class No.: 589.9/2 20

Advances in applied biotechnology series. Published/Created: Woodlands, Tex.: Portfolio Pub. Co.; Houston: Gulf Pub. Co., Book Division, c1989- Description: v.: ill.; 24 cm. Vol. 1- Current Frequency: Annual ISSN: 1053-4490 CODEN: AASEE6 Indx'd selectively by: Biological abstracts 0006-3169 1990- Chemical abstracts 0009-2258 Subjects: Biotechnology. Biotechnology--periodicals. LC Classification: CLASSED SEPARATELY Dewey Class No.: 660 12

Advances in applied lipid research. Published/Created: London; Greenwich, Conn.: JAI Press Ltd., c1992-c1996. Description: 2 v.: ill.; 24 cm. Vol. 1 (1992)-v. 2 (1996). Current Frequency:

Annual CODEN: AALREA Notes: SERBIB/SERLOC merged record Subjects: Lipids--Biotechnology--Periodicals. Lipids--Metabolism--Periodicals. Lipids--Synthesis--Periodicals. Lipids--analysis--periodicals. Lipids--metabolism--periodicals. Lipids--periodicals. LC Classification: TP248.65.L57 A38

Advances in bioprocess technology: industrial/specialty chemicals via biological sources/routes. Published/Created: Fort Lee, NJ: Technical Insights, c1985. Related Authors: Technical Insights, Inc. Description: viii, 265 leaves; 29 cm. ISBN: 0914993127 (pbk.) Notes: Includes bibliographies and indexes. Subjects: Biochemical engineering--Research. Biotechnology--Research. Biotechnology industries. Series: Emerging technologies; no. 14 LC Classification: TP248.3 .A383 1985 Dewey Class No.: 660/.63/072 19

Advances in biotechnology commercialization / International Resource Development Inc. Published/Created: Norwalk, Conn. (30 High St., Norwalk 06851): IRD, 1981. Related Authors: International Resource Development, inc. Description: xiv, 249 leaves: ill.; 28 cm. Subjects: Biotechnology industries. Genetic engineering industry. Market surveys. Series: Report (International Resource Development, inc.); #184. Variant Series: Report / International Resource Development Inc.; #184 LC Classification: HD9999.G452 A34 1981 Dewey Class No.: 380.1/456606 19

Advances in biotechnology for the manufacture of commodity & specialty chemicals. Published/Created: New York, N.Y.: Herwin International, c1999. Related Authors: Herwin International. Description: xiii, 216 p.: ill.; 28 cm. ISBN: 0471363502 Notes: "H-10"--Spine. Includes bibliographical references (p. 215-216). Subjects: Biotechnology. Biosynthesis. LC Classification: TP248.2 .A35 1999 Dewey Class No.: 660.6 21

Advances in gene technology. Published/Created: London; Greenwich, Conn.: JAI Press, 1990- Description: v.: ill.; 24 cm. Vol. 1 (1990)- Current Frequency: Annual Notes: SERBIB/SERLOC merged record Indx'd selectively by: Bibliography of agriculture 0006-1530 Subjects: Biotechnology--Periodicals. Genetic engineering--Periodicals. Biotechnology--periodicals. Genetic Engineering--periodicals. LC Classification: TP248.6 .A38

Advances in heat and mass transfer in biotechnology, 1996: presented at the 1996 International Mechanical Engineering Congress and Exposition, November 17-22, 1996, Atlanta, Georgia / sponsored by the Heat Transfer Division, ASME, the Bioengineering Division, ASME; edited by Linda J. Hayes, Scott Clegg.

Published/Created: New York, N.Y.: American Society of Mechanical Engineers, c1996. Related Authors: Hayes, Linda J. Clegg, Scott, 1956- American Society of Mechanical Engineers. Heat Transfer Division. American Society of Mechanical Engineers. Bioengineering Division. International Mechanical Engineering Congress and Exposition (1996: Atlanta, Ga.) Description: vi, 81 p.: ill.; 28 cm. ISBN: 0791815420 Contents: Processing of biomaterials -- Clinical therapies: advances in bioheat and mass transfer -- Nucleation crystallization and phase change in biological systems -- Advances in instrumentation and bioheat and mass transfer. Notes: Includes bibliographical references and index. Subjects: Biotechnology--Congresses. Biomedical engineering--Congresses. Heat--Transmission--Congresses. Mass transfer--Congresses. Biological systems--Congresses. Series: HTD (Series); vol. 337. BED (Series); vol. 34. Variant Series: HTD; vol. 337 BED; vol. 34 LC Classification: TP248.14 .A38 1996 Dewey Class No.: 660.6 21

Advances in heat and mass transfer in biotechnology, 1999: presented at the 1999 ASME International Mechanical Engineering Congress and Exposition, November 14-19, 1999, Nashville, Tennessee / sponsored by the Heat Transfer Division, ASME, the Bioengineering Division, ASME; edited by Elaine P. Scott. Published/Created: New York, N.Y.: American Society of Mechanical Engineers, c1999. Related Authors: Scott, Elaine P. American Society of Mechanical Engineers. Heat Transfer Division. American Society of Mechanical Engineers. Bioengineering Division. International Mechanical

Engineering Congress and Exposition (1999: Nashville, Tenn.) Description: vii, 251 p.: ill.; 28 cm. ISBN: 0791816435 Notes: Includes bibliographical references and index. Subjects: Heat--Transmission--Congresses. Mass transfer--Congresses. Biomass--Congresses. Biological systems--Congresses. Biotechnology--Congresses. Series: HTD (Series); vol. 363. BED (Series); vol. 44. Variant Series: HTD; vol. 363 BED; vol. 44 LC Classification: QC319.8 .I575 1999

Advances in heat and mass transfer in biotechnology--2001: presented at the 2001 ASME International Mechanical Engineering Congress and Exposition: November 11-16, 2001, New York, New York / sponsored by the Heat Transfer Division, ASME, the Bioengineering Division, ASME; edited by Elaine P. Scott, John C. Bischof. Parallel Heat and mass transfer in biotechnology, 2001 Published/Created: New York, New York: American Society of Mechanical Engineers, c2001. Related Authors: Scott, Elaine P. (Elaine Patricia), 1957- Bischof, John C. American Society of Mechanical Engineers. Heat Transfer Division. American Society of Mechanical Engineers. Bioengineering Division. International Mechanical Engineering Congress and Exposition (2001: New York, New York) Description: vi, 133 p.: ill.; 28 cm. ISBN: 0791835677 Notes: Includes bibliographic references and author index. Subjects: Heat--Transmission--Congresses. Mass transfer--Congresses. Biomass--Congresses. Biological systems--Congresses. Biotechnology--Congresses. Series: HTD (Series); vol. 370. BED (Series); vol. 52. Variant Series: HTD; vol. 370 BED; vol. 52

Advances in microbial biotechnology / edited by J.P. Tewari ... [et al.]. Published/Created: New Delhi: A.P.H. Pub. Corp., 1999. Related Authors: Tewari, J. P. Mukerji, K. G. Description: xlvii, 567 p.: ill.; 25 cm. ISBN: 8176480789 Summary: Published to honour Prof. Krishna Gopal Mukerji for his endeavour in mycology, plant pathology, microbial ecology and microbial biotechnology. Notes: Includes bibliographical references and index. Subjects: Microbial biotechnology. Industrial microbiology. Microbial ecology. LC Classification: TP248.27.M53 A384 1999 Dewey Class No.: 660.6 21

Advances in nucleic acid and protein analyses, manipulation, and sequencing: 26-27 January 2000, San Jose, California / Patrick A. Limbach ... [et al.], chairs/editors; sponsored by SPIE--the International Society for Optical Engineering [and] IBOS--the International Biomedical Optics Society. Published/Created: Bellingham, Wash., USA: SPIE, c2000. Related Authors: Limbach, Patrick A. Society of Photo-optical Instrumentation Engineers. International Biomedical Optics Society. Description: ix, 242 p.: ill. (some col.); 28 cm. ISBN: 0819435422 Notes: Includes bibliographical references and index. Subjects: Proteins--Analysis--Congresses. Nucleotide sequence--Congresses. Biotechnology--Congresses. Series: Progress in biomedical optics; vol. 1, no. 20 Proceedings of SPIE--the International Society for Optical Engineering; v. 3926. Variant Series: Proceedings of SPIE; v. 3926 LC Classification: TP248.65.P76 A375 2000 Dewey Class

No.: 660.6/3 21

Advances in plant cell biochemistry and biotechnology. Published/Created: London, England; Greenwich, Conn.: JAI Press, c1992- Description: v.: ill.; 24 cm. Vol. 1 (1992)- Current Frequency: Annual ISSN: 1071-9695 Cancel/Invalid LCCN: sn 93015545 CODEN: ACBBES Notes: SERBIB/SERLOC merged record Indexed by: Bibliography of agriculture 0006-1530 Indx'd selectively by: Chemical abstracts 0009-2258 Subjects: Plant biotechnology--Periodicals. Botanical chemistry--Periodicals. Biochemistry--periodicals. Biotechnology--periodicals. Plants--cytology--periodicals. LC Classification: TP248.27.P55 A36 Dewey Class No.: 581.19/2/05 20

Advances in protein design: international workshop 1988 / edited by H. Blöcker ... [et al.]. Published/Created: [Weinheim, Federal Republic of Germany]: VCH; New York, NY, USA: Distribution, VCH Publishers, c1989. Related Authors: Blöcker, H. (Helmut) Description: x, 217 p.: ill.; 24 cm. ISBN: 0895739534 (VCH Publishers) Notes: Includes bibliographical references. Subjects: Protein engineering--Congresses. Proteins--Biotechnology--Congresses. Proteins--Synthesis--Congresses. Series: GBF monographs, 0930-4320; v. 12 LC Classification: TP248.65.P76 A38 1989 Dewey Class No.: 660/.63 20

Aldridge, Susan. The thread of life: the story of genes and genetic engineering / Susan Aldridge. Published/Created: Cambridge [England]; New York, NY, USA: Cambridge University Press, 1996. Description: vii, 258 p.: ill.; 24 cm. ISBN: 0521465427 Notes: Includes bibliographical references (p. [253]-254) and index. Subjects: Genetic engineering. Genetics. DNA. Biotechnology. LC Classification: QH442 .A43 1996 Dewey Class No.: 575.1/0724 20

Algal and cyanobacterial biotechnology / editors, R.C. Cresswell, T.A.V. Rees, N. Shah. Published/Created: Harlow, Essex, England: Longman Scientific & Technical; New York: Wiley, 1989. Related Authors: Cresswell, R. C., 1956- Rees, T. A. V., 1953- Shah, N. (Nishith), 1957- Description: xvi, 341 p.: ill.; 25 cm. ISBN: 0470214775 (Wiley): Notes: Includes bibliographical references. Subjects: Algae--Biotechnology. Cyanobacteria--Biotechnology. LC Classification: TP248.27.A46 A44 1989 Dewey Class No.: 660/.62 20

Algal biomass technologies: an interdisciplinary perspective: proceedings of a workshop on the present status and future directions for biotechnologies based on algal biomass production, April 5-7, 1984, University of Colorado, Boulder / edited by William R. Barclay and Robins P. McIntosh. Published/Created: Berlin: J. Cramer, 1986. Related Authors: Barclay, William R. McIntosh, Robins P. University of Colorado, Boulder. Description: viii, 273 p.: ill.; 28 cm. ISBN: 3443510035 (pbk.) Notes: Includes bibliographies and index. Subjects: Algae--Biotechnology--Congresses. Series: Beihefte zur Nova Hedwigia; Heft 83 LC Classification: QK504 .N62 Heft 83 TP248.27.A46

Analytical biotechnology: capillary electrophoresis and chromatography /

Csaba Horváth, editor, John G. Nikelly, editor. Published/Created: Washington, DC: American Chemical Society, 1990. Related Authors: Horváth, Csaba, 1930- Nikelly, J. G. (John G.) American Chemical Society. Division of Analytical Chemistry. American Chemical Society. Meeting (196th: 1988: Los Angeles, Calif.) Description: x, 213 p.: ill.; 24 cm. ISBN: 0841218196 Notes: "Developed from a symposium sponsored by the Division of Analytical Chemistry at the 196th national Meeting of the American Chemical Society, Los Angeles, California, September 25-30, 1988." Includes bibliographical references and index. Subjects: Capillary electrophoresis--Congresses. Chromatographic analysis--Congresses. Proteins--Analysis--Congresses. Biotechnology--Technique--Congresses. Series: ACS symposium series, 0097-6156; 434 LC Classification: TP248.25.C37 A53 1990 Dewey Class No.: 660/.63 20

ASFA marine biotechnology abstracts. Published/Created: Bethesda, MD: Cambridge Scientific Abstracts, c1989- Related Authors: Cambridge Scientific Abstracts, inc. United Nations. Office for Ocean Affairs and the Law of the Sea. Food and Agriculture Organization of the United Nations. Intergovernmental Oceanographic Commission. Description: v.; 23 cm. Vol. 2, no. 1 (Mar. 1990)- Current Frequency: Quarterly Continues: Marine biotechnology abstracts 1043-8971 (OCoLC)19596692 (DLC) 90648488 ISSN: 1054-2027 Incorrect ISSN: 1043-8971 Notes: Compiled by: the United Nations Office for Ocean Affairs and the Law of the Sea, the Food and Agriculture Organization of the United Nations, and: the Intergovernmental Oceanographic Commission, with the collaboration of the ASFIS Input Partners. SERBIB/SERLOC merged record Subjects: Marine biotechnology--Abstracts--Periodicals. Biotechnology--abstracts. Genetic Engineering--abstracts. Marine Biology--abstracts. Series: Aquatic sciences & fisheries abstracts series LC Classification: TP248.27.M37 M37 Dewey Class No.: 660/.6 20

Aspergillus / edited by J.E. Smith. Published/Created: New York: Plenum Press, c1994. Related Authors: Smith, John E. Description: xv, 273 p.: ill.; 24 cm. ISBN: 030644545X Notes: Includes bibliographical references and indexes. Subjects: Aspergillus--Biotechnology. Fungi--Biotechnology. Series: Biotechnology handbooks; v. 7 The Language of science LC Classification: TP248.27.F86 A873 1994 Dewey Class No.: 660/.62 20

Aspergillus: 50 years on / edited by S.D. Martinelli and J.R. Kinghorn. Published/Created: Amsterdam; New York: Elsevier, 1994. Related Authors: Martinelli, S. D. (Sylvia D) Kinghorn, James R. Description: xxv, 851 p.: ill.; 25 cm. ISBN: 044481762X: Notes: Includes bibliographical references and index. Subjects: Aspergillus--Genetics. Aspergillus nidulans--Genetics. Aspergillus--Metabolism. Aspergillus nidulans--Metabolism. Aspergillus--Biotechnology. Series: Progress in industrial microbiology; v. 29 LC Classification: QH470.A85 A86 1994 Dewey Class No.: 589.2/3 20

Aspergillus: biology and industrial applications / edited by J.W. Bennett,

M.A. Klich. Published/Created: Boston: Butterworth-Heinemann, c1992. Related Authors: Bennett, J. W. Klich, Maren A. Description: xvi, 416 p.: ill.; 25 cm. ISBN: 0750691247 (alk. paper) Notes: Includes bibliographical references and index. Subjects: Aspergillus--Biotechnology. Series: Biotechnology (Reading, Mass.); 23. Variant Series: Biotechnology series; 23 LC Classification: TP248.27.F86 A87 1992 Dewey Class No.: 589.2/3 20

Assessing ecological risks of biotechnology / edited by Lev R. Ginzburg. Published/Created: Boston: Butterworth-Heinemann, c1991. Related Authors: Ginzburg, Lev R. Description: xvi, 379 p.: ill.; 24 cm. ISBN: 0409901997 Notes: Includes bibliographical references and index. Subjects: Microbial biotechnology--Safety measures. Microbial biotechnology--Environmental aspects. Genetic engineering--Safety measures. Ecology. Series: Biotechnology (Reading, Mass.); 15. Variant Series: Biotechnology series; 15 LC Classification: TP248.27.M53 A87 1991 Dewey Class No.: 660/.62/0289 20

ATCC microbes & cells at work: an index to ATCC strains with special applications / edited by M.J. Edwards ... [et al.]. Edition Information: 1st ed. Published/Created: Rockville, Md.: American Type Culture Collection, 1988. Related Authors: Edwards, M. J. (Malcolm John) American Type Culture Collection. Related Titles: ATCC microbes and cells at work. Microbes & cells at work. Microbes and cells at work. Description: v, 225 p.; 28 cm. ISBN: 0930009215 Notes: Cover ATCC microbes and cells at work.

Bibliography: p. 116-225. Subjects: American Type Culture Collection--Catalogs. Biomolecules--Synthesis--Bibliography--Indexes. Microorganisms--Biotechnology--Catalogs and collections. Animal cell biotechnology--Catalogs and collections. LC Classification: Z7914.B32 A86 1988 TP248.2 Dewey Class No.: 660/.6/0294 20

Bacillus molecular genetics and biotechnology applications / edited by A.T. Ganesan, James A. Hoch. Published/Created: Orlando, Fla.: Academic Press, 1986. Related Authors: Ganesan, A. T. Hoch, James A. International Conference on the Genetics and Biotechnology of Bacilli (3rd: 1985: Stanford University) Description: xi, 497 p.: ill.; 24 cm. ISBN: 0122741552 (alk. paper) Notes: Proceedings of the Third International Conference on the Genetics and Biotechnology of Bacilli, held at Stanford University, July 15-17, 1985. Includes bibliographies and index. Subjects: Bacillus (Bacteria)--Genetics--Congresses. Bacillus (Bacteria)--Biotechnology--Congresses. Bacillus subtilis--Genetics--Congresses. Bacillus subtilis--Biotechnology--Congresses. LC Classification: QR82.B3 B33 1986 Dewey Class No.: 589.9/5 19

Bacillus subtilis: molecular biology and industrial application / edited by Bunji Maruo, Hiroshi Yoshikawa. Published/Created: Tokyo: Kodansha; Amsterdam; New York: Elsevier; New York, NY: Exclusive sales for the U.S. and Canada, Elsevier Science Pub. Co., 1989. Related Authors: Maruo, Bunji, 1917- Yoshikawa, Hiroshi, 1933- Description: xv, 267 p.: ill.; 24 cm. ISBN: 0444988521 (U.S.) Notes:

Includes bibliographical references. Subjects: Bacillus subtilis-- Biotechnology. Molecular microbiology. Series: Topics in secondary metabolism; 1 LC Classification: QR82.B3 B334 1989 Dewey Class No.: 589.9/5 20

Bacteria / edited by L.R. Hill and B.E. Kirsop. Published/Created: Cambridge [England]; New York: Cambridge University Press, 1991. Related Authors: Hill, L. R. (Leslie Rowland), 1935- Kirsop, B. E. Description: xiv, 186 p.: ill.; 24 cm. ISBN: 052135224X Notes: Includes bibliographical references and index. Subjects: Bacteria--Type specimens--Catalogs and collections. Bacteria--Type specimens--Catalogs and collections Directories. Biotechnological microorganisms--Catalogs and collections Directories. Bacteria. Biotechnology. Series: Living resources for biotechnology LC Classification: QR64.5 .B33 1990 Dewey Class No.: 660/.62 20

Bajpai, P. (Pratima) Biotechnology for environmental protection in the pulp and paper industry / P. Bajpai, P.K. Bajpai, R. Kondo. Published/Created: Berlin; New York: Springer, c1999. Related Authors: Bajpai, P. K. (Pramod Kumar) Kondo, R. (Ryuichiro), 1949- Description: 266 p.: ill.; 25 cm. ISBN: 3540656774 (acid-free paper) Notes: Includes bibliographical references. Subjects: Wood-pulp--Biotechnology. Wood-pulp industry--Environmental aspects. Paper industry--Environmental aspects. Paper chemistry. LC Classification: TS1176.6.B56 B35 1999 Dewey Class No.: 676/.042 21

Bioactive microbial products 3: downstream processing / edited by J.D. Stowell, P.J. Bailey, and D.J. Winstanley. Published/Created: London; Orlando: Published for the Society for General Microbiology by Academic Press, 1986. Related Authors: Stowell, J. D. Bailey, P. J. Winstanley, D. J. Society for General Microbiology. Fermentation Group. Related Titles: Bioactive microbial products three. Description: x, 242 p.: ill.; 24 cm. ISBN: 0126729603 0126729611 (pbk.) Notes: Based on a symposium of the Fermentation Group of the Society for General Microbiology, held at the University of Kent at Canterbury and Pfizer Limited, Sandwich. Includes bibliographical references. Subjects: Microbial biotechnology--Congresses. Microbial products--Congresses. Series: Special publications of the Society for General Microbiology; 18 LC Classification: TP248.27.M53 B56 1986 Dewey Class No.: 660/.62 20

Biocatalysis and biomimetics / James D. Burrington, editor, Douglas S. Clark, editor. Published/Created: Washington, DC: American Chemical Society, 1989. Related Authors: Burrington, James D., 1951- Clark, Douglas S., 1957- American Chemical Society. Division of Petroleum Chemistry. American Chemical Society. Division of Industrial and Engineering Chemistry. American Chemical Society. Biotechnology Secretariat. American Chemical Society. Meeting (195th: 1988: Toronto, Ont.) Chemical Congress of North America (3rd: 1988: Toronto, Ont.) Description: xiii, 169 p.: ill. (some col.); 24 cm. ISBN: 0841216118 Notes: "Developed from a symposium sponsored by the Divisions of Petroleum Chemistry, Inc., and of

Industrial and Engineering Chemistry, Inc., as part of the program of the Biotechnology Secretariat at the Third Chemical Congress of North America (195th National Meeting of the American Chemical Society), Toronto, Ontario, Canada, June 5-11, 1988." Includes bibliographies and indexes. Subjects: Enzymes--Biotechnology--Congresses. Biomimetics--Biotechnology--Congresses. Series: ACS symposium series, 0097-6156; 392 LC Classification: TP248.65.E59 B56 1989 Dewey Class No.: 660.2/995 19

Biocatalysis and biotransformation. Published/Created: [Switzerland?]: Harwood Academic Publishers, c1995- Related Authors: European Federation of Biotechnology. Working Party on Applied Biocatalysis. Description: v.: ill.; 25 cm. Vol. 12, no. 1 (Feb. 1995)- Current Frequency: Four no. a year Continues: Biocatalysis 0886-4454 (DLC) 87643338 (OCoLC)12890509 ISSN: 1024-2422 Cancel/Invalid LCCN: sn 95015411 CODEN: BOBOEQ Notes: Title from cover. Published in association with the Workin

Biocatalysis. Published/Created: Chur [Switzerland]; New York: Harwood Academic Publishers, c1987-c1994. Related Authors: European Federation of Biotechnology. Working Party on Applied Biocatalysis. Description: 11 v.: ill.; 25 cm. Vol. 1, no. 1 (May 1987)- v. 11, no. 4 Current Frequency: Quarterly Continued by: Biocatalysis and biotransformation 1024-2422 (DLC) 97661024 (OCoLC)32474843 ISSN: 0886-4454 Cancel/Invalid LCCN: sn 85003420 CODEN: BIOCED Notes: Title from cover. Distributed: Berkshire, UK: STBS Ltd., <1992-

Published in association with the Working Party on Applied Biocatalysis of the European Federation of Biotechnology, <1992- SERBIB/SERLOC merged record Additional Form Avail.: Also available in microform. Subjects: Enzymes--Biotechnology--Periodicals. Enzymes--Industrial applications--Periodicals. Biotechnology--periodicals. Catalysis--periodicals. Enzymes--periodicals. LC Classification: TP248.65.E59 B54 Dewey Class No.: 660/.634/05 20

Biomass energy systems: proceedings of the International Conference, 26-27 February 1996, New Delhi / edited by Venkata Ramana, P., Srinivas, S.N.; foreword R.K. Pachauri; [jointly organised by] British Council Division, British High Commission, TERI. Published/Created: New Delhi: Tata Energy Research Institute, c1997. Related Authors: Venkata Ramana, P. Srinivas, S. N. Great Britain. High Commission (India). British Council Division (New Delhi, India) Tata Energy Research Institute. International Conference on Biomass Energy Systems (1996: New Delhi, India) Description: xv, 478 p.: ill.; 25 cm. ISBN: 8185419256 Notes: "International Conference on Biomass Energy Systems held on 26-27 February 1996 at New Delhi"--Pref. Includes bibliographical references. Subjects: Biomass energy sources--Developing countries--Congresses. Renewable energy sources--Developing countries Congresses. Biotechnology industries--Developing countries Congresses. LC Classification: TP360 .B5894 1997 Dewey Class No.: 333.95/39/0971724 21

Biomining: theory, microbes, and industrial processes / [edited by] Douglas E.

Rawlings. Published/Created: Georgetown, Tex.: Landes Bioscience, 1997. Projected Pub. Date: ---- Related Authors: Rawlings, Douglas E. Description: p. cm. ISBN: 1570594708 (alk. paper) Notes: Includes bibliographical references and index. Subjects: Minerals--Biotechnology. Bacterial leaching. Bioremediation. Extreme environments--Microbiology. Thermophilic bacteria. Series: Biotechnology intelligence unit (Unnumbered) Variant Series: Biotechnology intelligence unit LC Classification: TN688.3.B33 B55 1997 Dewey Class No.: 622/.7 21

Biomolecular engineering in the European community: achievements of the research programme (1982-1986): final report / edited by E. Magnien. Published/Created: Dordrecht; Boston: M. Nijhoff for the Commission of the European Communities; Norwell, MA, USA: Distributors for the U.S. and Canada, Kluwer Academic Publishers, 1986. Related Authors: Magnien, E. Commission of the European Communities. Description: iv, 1172 p.: ill.; 25 cm. ISBN: 9024734002 Notes: Includes bibliographies and indexes. Subjects: Biochemical engineering--Research--Europe. Biotechnology--Research--Europe. Genetic engineering--Research--Europe. Series: EUR (Series); 10658 EN. Variant Series: EUR; 10658 EN LC Classification: TP248.3 .B57 1986 Dewey Class No.: 660/.63 19

Biomolecular self-assembling materials: scientific and technological frontiers / Panel on Biomolecular Materials, Solid States Sciences Committee, Board on Physics and Astronomy, Commission on Physical Sciences, Mathematics, and Applications, National Research Council. Published/Created: Washington, D.C.: National Academy Press, 1996. Related Authors: National Research Council (U.S.). Panel on Biomolecular Materials. Description: ix, 32 p.: ill.; 28 cm. Notes: Includes bibliographical references. Subjects: Biomolecules. Molecular biology. Materials--Biotechnology. Series: Compass series (Washington, D.C.) Variant Series: The Compass series LC Classification: QH505 .B4685 1996 Dewey Class No.: 572.8 21

Biomolecules in organic solvents / editor, Armando Gómez-Puyou; associate editors, Alberto Darszon, Marietta Tuena de Gómez-Puyou. Published/Created: Boca Raton: CRC Press, c1992. Related Authors: Gómez-Puyou, Armando. Darszon, Alberto. Tuena de Gómez-Puyou, Marietta. Description: 266 p.: ill.; 26 cm. ISBN: 0849348234 Notes: Includes bibliographical references and index. Subjects: Enzymes--Biotechnology. Organic solvents--Biotechnology. Reversed micelles. Enzymes--Solubility. LC Classification: TP248.65.E59 B58 1992 Dewey Class No.: 661/.807 20

Bio-organic compounds chemistry and biomedical applications / editors, Milton Fingerman, Rachakonda Nagablushanam. Published/Created: Enfield, (NH): Science Publishers, 2001. Projected Pub. Date: 0109 Related Authors: Fingerman, Milton, 1928- Nagabhushanam, Rachakonda. Description: p. cm. ISBN: 1578081351 Notes: Includes bibliographical references and index. Subjects: Marine biotechnology. Marine pharmacology. Bioorganic chemistry. Series: Recent advances in marine biotechnology; v. 6 LC Classification: TP248.27.M37 B546 2001 Dewey Class No.: 660.6 21

Biopolymers from renewable resources / D.L. Kaplan (ed.). Published/Created: Berlin; New York: Springer, c1998. Related Authors: Kaplan, David, 1953- Description: xviii, 417 p.: ill.; 24 cm. ISBN: 354063567X (acid-free paper) Notes: Includes bibliographical references and index. Subjects: Polymers--Biotechnology. Series: Macromolecular systems, materials approach LC Classification: TP248.65.P62 B55 1998 Dewey Class No.: 668.9 21

Biopolymers/natural polymers: the technology is ready, the markets are waiting, a host of opportunities can now be exploited. Published/Created: Englewood/Fort Lee, NJ: Technical Insights, c1994. Related Authors: Technical Insights, Inc. Description: 145 leaves; 30 cm. ISBN: 1562170082 Notes: "R-198"--t.p. verso. Subjects: Biotechnology--Economic aspects-- United States. Biopolymers. Macromolecules. Polymers. Market surveys--United States. LC Classification: TP248.185 .B55 1994 Dewey Class No.: 380.1/456684 21

Biopolymers: sophisticated materials with growing market potential. Edition Information: 2nd ed. Published/Created: Englewood, NJ: Technical Insights, c1998. Related Authors: Technical Insights, Inc. Description: vii, 107 p.; 28 cm. ISBN: 0471314226 Notes: "R-230"--t.p. verso. Subjects: Biotechnology--Economic aspects--United States. Biopolymers. Macromolecules. Polymers. Market surveys--United States. LC

Classification: TP248.185 .B55 1998 Dewey Class No.: 338.4/76606/3 21

Biopolymers: utilizing nature's advanced materials / Syed H. Imam, editor, Richard V. Greene, editor, Baqar R. Zaidi, editor. Published/Created: Washington, DC: American Chemical Society; [New York]: Distributed by Oxford University Press, c1999. Related Authors: Imam, Syed H. Greene, Richard V. Zaidi, Baqar R. Chemical Congress of North America (5th: 1997: Cancún, Mexico) Description: xii, 305 p.: ill.; 24 cm. ISBN: 0841236070 (cloth: alk. paper) Notes: "Developed from a symposium at the Fifth Chemical Congress of North America, Cancun, Quintana Roo, Mexico, November 11- 15, 1997." Includes bibliographical references and indexes. Subjects: Polysaccharides--Biotechnology-- Congresses. Biopolymers-- Biotechnology--Congresses. Series: ACS symposium series, 0097-6156; 723 LC Classification: TP248.65.P64 B56 1999 Dewey Class No.: 660.6/3 21

Bioprocess computations in biotechnology / editor, T.K. Ghose. Published/Created: New York: E. Horwood, 1990- Related Authors: Ghose, T. K. Description: v. <1: ill.; 24 cm. ISBN: 0130846740 (lib. ed.: v. 1) 0130846589 (student ed.: v. 1) Notes: Includes index. Subjects: Biotechnology--Mathematics. Biotechnology--Problems, exercises, etc. Biochemical engineering-- Mathematics. Biochemical engineering- -Problems, exercises, etc. Series: Ellis Horwood books in the biological sciences. Series in biochemistry and biotechnology. Variant Series: Ellis Horwood series in biochemistry and biotechnology LC Classification: TP248.24 .B56 1990 Dewey Class No.:

660/.6 20 -- Library Holds: 1

BioProcess engineering. Published/Created: Tampa, FL: U.S. Bureau of Pharmaceutical Research, 1984- Related Authors: Bioprocess Engineering Society International. Description: v.: ill.; 28 cm. Vol. 1, no. 1 also called "premier issue". Vol. 1, no. 1- Current Frequency: Monthly ISSN: 0883-0878 Cancel/Invalid LCCN: sn 85007503 Notes: Title from cover. Published: Dunedin, Fla.: Media Division, Apr. 1985- Official journal of the: Bioprocess Engineering Society International, Apr. 1985- SERBIB/SERLOC merged record Subjects: Biochemical engineering-- Periodicals. Biotechnology--Periodicals. LC Classification: TP248.3 .B5585 Dewey Class No.: 660/.63 20

Bioprocess engineering: the first generation / editor, Tarun K. Ghose; special adviser, Mike Winkler. Published/Created: Chichester: E. Horwood; New York, NY: Distributored in USA and Canada by Chapman and Hall, 1989. Related Authors: Ghose, T. K. Description: 389 p.: ill.; 25 cm. ISBN: 0745807097 (E. Horwood) 0412023318 (Routledge, Chapman and Hall) Notes: Includes bibliographical references. Subjects: Biochemical engineering. Biotechnology. Series: Ellis Horwood series in biochemistry and biotechnology LC Classification: TP248.3 .B5874 1989 Dewey Class No.: 660/.6 20

Bioprocess production of flavor, fragrance, and color ingredients / edited by Alan Gabelman. Published/Created: New York: Wiley, c1994. Related Authors: Gabelman, Alan. Description: xiii, 361 p.: ill.; 25 cm. ISBN: 0471038210 (acid-free paper) Notes: "A Wiley-Interscience publication." Includes bibliographical references and index. Subjects: Flavoring essences-- Biotechnology. LC Classification: TP418 .B565 1994 Dewey Class No.: 664/.5 20

Bioprocess technology. Published/Created: New York: Dekker, c1986-c1999. Description: 24 v.: ill.; 24 cm. Vol. 1-24. Continued by: Biotechnology and bioprocessing series (DLC) 00220375 (OCoLC)44018718 ISSN: 0888-7470 CODEN: BPTEEP Indx'd selectively by: Chemical abstracts 0009-2258 1986- BIOTECHSEEK 1990- Subjects: Biotechnology. Genetic engineering. Biomedical Engineering--periodicals. Technology, Medical--periodicals. LC Classification: NOT IN LC Dewey Class No.: 660 11

Bioprocessing and biotreatment of coal / edited by Donald L. Wise. Published/Created: New York: M. Dekker, c1990. Related Authors: Wise, Donald L. (Donald Lee), 1937- Description: xiii, 744 p.: ill.; 26 cm. ISBN: 0824783050 (alk. paper) Notes: Includes bibliographical references and index. Subjects: Coal--Cleaning. Coal-- Biotechnology. LC Classification: TP325 .B475 1990 Dewey Class No.: 662.6/23 20

Bioprocessing technology. Published/Created: Englewood, NJ: Technical Insights, Inc., c1985-c1991. Related Authors: Technical Insights, Inc. Description: 7 v.; 28 cm. Vol. 7, no. 1 (Jan. 1985)-vol. 13, no. 5 (May 1991). Current Frequency: Monthly Continues: Biomass digest 0163-6766 (OCoLC)4447555 (DLC)sn 78006842

Continued by: Industrial bioprocessing 1056-7194 (DLC)sn 91004360 (OCoLC)23823201 ISSN: 0885-5625 Incorrect ISSN: 0163-6766 CODEN: BITEEA Notes: Title from caption. Indx'd selectively by: Abstract bulletin of the Institute of Paper Chemistry 0020-3033 Bibliography of agriculture 0006-1530 Subjects: Biomass energy--Periodicals. Natural products--Industrial applications--Periodicals. Power resources--Periodicals. Biological Products--periodicals. Biotechnology--periodicals. LC Classification: CURRENT ISSUES ONLY Dewey Class No.: 660 11

Bioproducts processing: technologies for the tropics / edited by M.A. Hashim. Published/Created: Rugby, UK: Institution of Chemical Engineers, c1994. Related Authors: Hashim, M. A. Institute of Advanced Studies (Universiti Malaya) Institution of Engineers, Malaysia. Institution of Chemical Engineers (Great Britain) International Symposium on Bioproducts Processing (1994: Kuala Lumpur, Malaysia) Description: x, 246 p.: ill.; 22 cm. ISBN: 0852953305 Notes: "A collection of selected papers presented at the International Symposium on Bioproducts Processing in January 1994"--P. v. "A four day symposium organised by the Institute of Advanced Studies (University of Malaya) with the co-operation of the Institution of Chemical Engineers and the Institute of Engineers Malaysia and held at the Institute of Advanced Studies, Kuala Lumpur, Malaysia 4-7 January 1994"--P. iii. Errata replacement pages included for p. vii and ix. Includes bibliographical references and index. Subjects: Biological products--Tropics--

Congresses. Biochemical engineering--Methodology--Congresses. Biotechnology--Methodology--Congresses. Series: Symposium series (Institution of Chemical Engineers (Great Britain)); no. 137. Variant Series: Symposium series / Institution of Chemical Engineers; no. 137 LC Classification: TP248.14 .B554 1994 Dewey Class No.: 660/.6/0913 20

Bioreactor immobilized enzymes and cells: fundamentals and applications / edited by Murray Moo-Young. Published/Created: London; New York: Elsevier Applied Science, c1988. Related Authors: Moo-Young, Murray. Description: xvi, 327 p.: ill.; 23 cm. ISBN: 1851661603 Notes: Includes bibliographies. Subjects: Immobilized cells--Industrial applications. Immobilized enzymes--Industrial applications. Biotechnology. Enzymes, Immobilized. LC Classification: TP248.25.I55 B56 1988 Dewey Class No.: 660/.62 19

Biorelated polymers: sustainable polymer science and technology / edited by Emo Chiellini ... [et al.]. Published/Created: New York: Kluwer Academic/Plenum, c2001. Related Authors: Chiellini, Emo. International Centre of Biopolymer Technology. International Conference on Biopolymer Technology (1st: 1999: Coimbra, Portugal) International Conference on Biopolymer Technology (2nd: 2000: Ischia, Italy) Description: xxiii, 391 p.: ill.; 26 cm. ISBN: 030646652X Notes: "Combined proceedings of the first and second International Conference on Biopolymer Technology, organised by the International Centre of Biopolymer Technology, held in Coimbra, Portugal on September 29-October 1, 1999 and in Ischia (Naples), Italy on October 25-27, 2000"--T.p. verso. Includes bibliographical references and index. Subjects: Biopolymers--Biotechnology. Polymers--Biodegradation. LC Classification: TP248.65.P62 B556 2001 Dewey Class No.: 668.9 21

Bioscience, biotechnology, and biochemistry. Published/Created: Tokyo, Japan: Japan Society for Bioscience, Biotechnology, and Agrochemistry, 1992- Related Authors: Nihon N⁻ogei Kagakkai. Description: v.: ill.; 29 cm. Vol. 56, no. 1 (Jan. 1992)- Current Frequency: Monthly Continues: Agricultural and biological chemistry 0002-1369 (DLC) 65058099 (OCoLC)5458841 ISSN: 0916-8451 Cancel/Invalid LCCN: sn 92037097 CODEN: BBBIEJ Notes: Title from cover. SERBIB/SERLOC merged record Indx'd selectively by: Index medicus 0019-3879 v62n2,Apr. 1998- BIOTECHSEEK 1992- Subjects: Biochemistry--Periodicals. Chemistry, Organic--Periodicals. Biotechnology--Periodicals. Biochemistry--periodicals. Biological Sciences--periodicals. Biotechnology--periodicals. LC Classification: QH345 .A35

Biosurfactants and biotechnology / edited by Naim Kosaric, W.L. Cairns, Neil C.C. Gray. Published/Created: New York: M. Dekker, c1987. Related Authors: Kosaric, Naim, 1928- Cairns, W. L. (William L.), 1942- Gray, Neil C. C., 1954- Description: viii, 342 p.: ill.; 24 cm. ISBN: 0824776798 Notes: Includes bibliographies and index. Subjects: Biosurfactants. Microbial surfactants. Biotechnology. Oil spills. Series: Surfactant science series; v. 25 LC Classification: TP248.B57 B56 1987

Dewey Class No.: 668/.14 19

Biotec: rivista bimestrale di biotecnologia. Published/Created: Brescia: CLAS, [1986?- Description: v.: ill. (chiefly col.); 28 cm. Vol. 1, n. 1 (genn.-febbr. 1986)- Ceased in 1994? Current Frequency: Bimonthly Continued by: BioTec (Genoa, Italy) (DLC)sn 99039767 (OCoLC)41984909 ISSN: 0393-9146 CODEN: BBRCED Notes: Title from cover. Chiefly in Italian; some issues include parallel text in English. SERBIB/SERLOC merged record Indx'd selectively by: Chemical abstracts 0009-2258 Subjects: Biotechnology--Periodicals. LC Classification: TP248.13 B558

Biotech buyers' guide / American Chemical Society. Published/Created: Washington, DC: American Chemical Society, c1990- Related Authors: American Chemical Society. Description: v.: ill.; 28 cm. Vol. 1 (1990)- Ceased in 1994. Cf. Letter from publisher. Current Frequency: Annual ISSN: 1067-2818 Cancel/Invalid LCCN: sn 90025281 CODEN: BBGUEC Notes: Title from cover. SERBIB/SERLOC merged record Subjects: Biological laboratory equipment industry--Directories. Biotechnology laboratories--Equipment and supplies Periodicals. Biotechnology--directories. Equipment and Supplies--directories. LC Classification: HD9706.65.B55 B56 Dewey Class No.: 660 12

Biotech daily. Published/Created: Washington, D.C.: King Pub. Group, c1992- Description: v.; 28 cm. Vol. 1, no. 1 (Aug. 11, 1992)- Current Frequency: Daily (Monday through Friday) Continues: Genetic engineering letter 0276-1882 (DLC) 85641462 (OCoLC)7332306 ISSN: 1067-1196 CODEN: BIDAEK Notes: Title from caption. SERBIB/SERLOC merged record Merged with: Pharmaceutical daily, to form: Pharmaceutical & biotech daily. Indx'd selectively by: Bibliography of agriculture 0006-1530 Subjects: Biotechnology--Periodicals. Biotechnology industries--Periodicals. Genetic engineering--Periodicals. Biotechnology--periodicals. Genetic Engineering--periodicals. LC Classification: TP248.13 .B5583 Dewey Class No.: 660/.6/05

Biotechnic & histochemistry: official publication of the Biological Stain Commission. Published/Created: Baltimore, MD: Williams & Wilkins, 1991- Related Authors: Biological Stain Commission. Description: v.: ill.; 28 cm. First 2 no. incorrectly called v. 1, no. 1-2, but constitute v. 66, no. 1-2. Vol. 1 [i.e. 66], no. 1-

Biotechnology & biological frontiers / edited by Philip H. Abelson. Published/Created: Washington, D.C.: American Association for the Advancement of Science, c1984. Related Authors: Abelson, Philip Hauge. Related Titles: Biotechnology and biological frontiers. Description: vi, 516 p., [7] p. of plates: ill. (some col.); 25 cm. ISBN: 0871682664 (pbk.) 0871683083 (hard): Notes: Includes bibliographies and index. Subjects: Biotechnology. Biology--Research. Series: AAAS publication; no. 84-8 LC Classification: Q181.A1 A68 no. 84-8 TP248.2 Dewey Class No.: 660/.62 19

Biotechnology & genetic engineering reviews. Published/Created: Newcastle upon Tyne: Intercept, c1984-

Description: v.: ill.; 24 cm. Vol. 1- ISSN: 0264-8725 Cancel/Invalid LCCN: sc 84001876 CODEN: BGERES Notes: SERBIB/SERLOC merged record Indx'd selectively by: Chemical abstracts 0009-2258 1984- Index Medicus 0019-3879 v1, 1984- Subjects: Biotechnology--Periodicals. Genetic engineering--Periodicals. Biochemical engineering--Periodicals. Biomedical Engineering--periodicals. Environmental Microbiology--periodicals. Genetic Engineering--periodicals. Microbiological Techniques--periodicals. Technology--periodicals. LC Classification: TP248.13 .B56 Dewey Class No.: 660/.6/05 19

Biotechnology / edited by Lynn Messina. Published/Created: New York: H.W. Wilson, 2000. Related Authors: Messina, Lynn. Description: ix, 186 p.; 26 cm. ISBN: 0824209850 (pbk.) Notes: Includes bibliographical references (p. 167-179) and index. Subjects: Biotechnology--Popular works. Genetic engineering--Popular works. Series: The Reference shelf; v. 72, no. 4 LC Classification: TP248.215 .B56 2000 Dewey Class No.: 660.6 21

Biotechnology '84: proceedings of a conference held in the Royal Irish Academy, 1-2 May, 1984 / edited by J. P. Arbuthnott. Published/Created: Dublin: The Academy, 1985. Related Authors: Arbuthnott, J. P. Royal Irish Academy. Ireland. National Board for Science and Technology. Society for General Microbiology. Irish Branch. Irish National Committee for Biology. Irish National Commission for Microbiology. Description: xii, 176 p.: ill.; 29 cm. ISBN: 0901714399 (pbk.) Notes: "Sponsored by the Royal Irish

Academy, National Board for Science and Technology, [and] Society for General Microbiology (Irish Branch), arranged under the auspices of the Irish National Committee for Biology and the Irish National Commission for Microbiology"--P. iii. Includes bibliographies. Subjects: Biotechnology--Congresses. LC Classification: TP248.14 .B559 1985 Dewey Class No.: 660/.6 20

Biotechnology and applied biochemistry. Published/Created: San Diego: Published for the International Union of Biochemistry by Academic Press, [c1986- Related Authors: International Union of Biochemistry. International Union of Biochemistry and Molecular Biology. Description: v.: ill.; 26 cm. Vol. 8, no. 1 (Feb. 1986)- Current Frequency: Bimonthly Continues: Journal of applied biochemistry 0161-7354 (DLC) 79643454 (OCoLC)4015541 ISSN: 0885-4513 Cancel/Invalid LCCN: sn 85005631 CODEN: BABIEC Notes: Title from cover. Imprint: London: Published for the International Union of Biochemistry and Molecular Biology by Portland Press, <1993- SERBIB/SERLOC merged record Indx'd selectively by: Abstract bulletin of the Institute of Paper Chemistry 0020-3033 1986- Bibliography of agriculture 0006-1530 1986- Biological abstracts 0006-3169 1986- Chemical abstracts 0009-2258 1986- Coal abstracts 0309-4979 1986- Energy research abstracts 0160-3604 1986- Excerpta medica 1986- Index medicus 0019-3879 1986- Life sciences collection 1986- Additional Form Avail.: Beginning with June 1998, full text also available to subscribers, in HTML or PDF, online via the World Wide Web; tables of contents and abstracts freely available beginning with 1987. Subjects: Biotechnology--Periodicals. Biochemical engineering--Periodicals. Biochemistry--Periodicals. Biochemistry--periodicals. Genetic Techniques--periodicals. Microbiological Techniques--periodicals. LC Classification: QP501 .J66

Biotechnology and bioactive polymers / edited by Charles G. Gebelein and Charles E. Carraher. Published/Created: New York: Plenum Press, c1994. Related Authors: Gebelein, Charles G. Carraher, Charles E. American Chemical Society. Symposium (1992: San Francisco, Calif.) Description: ix, 342 p.: ill.; 26 cm. ISBN: 0306446294 Notes: "Proceedings of an American Chemical Society Symposium on Biotechnology and Bioactive Polymers, held April 5-10, 1992, in San Francisco, California"--T.p. verso. Includes bibliographical references and index. Subjects: Biopolymers--Biotechnology--Congresses. LC Classification: TP248.65.P62 B565 1994 Dewey Class No.: 610/.28 20

Biotechnology and bioapplications of colloidal gold / compiled by Ralph M. Albrecht and Gisele M. Hodges. Published/Created: Chicago, IL., U.S.A.: Scanning Microscopy International, c1988. Related Authors: Albrecht, R. M. Hodges, Gisele M. Description: 312, viii p.: ill.; 28 cm. ISBN: 0931288398 Notes: Includes bibliographies and indexes. Subjects: Immunogold labeling. Biotechnology--Technique. Colloidal gold. LC Classification: QR187.I482 B56 1988 Dewey Class No.: 578 20

96   Biology

Biotechnology and bioengineering. Published/Created: New York: Wiley [etc.] Description: v. ill. 23 cm. v. 4- Mar. 1962- Current Frequency: Semimonthly with additional issues to make 4 v. of 7 issues per year, 2002- Former Frequency: Frequency varies, 1962-2001 Continues: Journal of biochemical and microbiological technology and engineering (OCoLC)2262998 (DLC)sf 84000014 ISSN: 0006-3592 Cancel/Invalid LCCN: sc 78002085 sn 78004263 Notes: SERBIB/SERLOC merged record Additional Form Avail.: Available also by subscription via the World Wide Web. Subjects: Biotechnology--Periodicals. Bioengineering--Periodicals. Biochemistry--periodicals. Microbiology--periodicals. Technology--periodicals. LC Classification: QH324 .B5 Dewey Class No.: 574 11

Biotechnology and environmental science: molecular approaches / edited by S. Mongkolsuk, P.S. Lovett, and J.E. Trempy. Published/Created: New York: Plenum Press, c1992. Related Authors: Mongkolsuk, S. Lovett, P. S. Trempy, J. E. Description: xi, 230 p.: ill.; 26 cm. ISBN: 030644352X Notes: "Proceedings of an International Conference on Biotechnology and Environmental Science: Molecular Approaches (BESMA), held August 21-24, 1990, in Bangkok, Thailand"--T.p. verso. Includes bibliographical references and index. Subjects: Biotechnology--Congresses. Bioremediation--Congresses. Agricultural biotechnology--Congresses. Microbial biotechnology--Congresses. Series: The Language of science LC Classification: TP248.14 .B5592 1992 Dewey Class No.: 660/.6

20

Biotechnology and farmer's rights: opportunities and threats for small-scale farmers in developing countries / edited by Hans Brouwer, Erik M. Stokhof, and Joske F.G. Bunders. Published/Created: Amsterdam: VU University Press, 1992. Related Authors: Brouwer, Hans. Stokhof, Erik M. Bunders, Joske F. G. Description: 122 p.; 24 cm. ISBN: 9053831932 Notes: "In 1991, the Netherlands Organisation for International Development Cooperation (NOVIB) and the Department of Biology and Society of the Vrije Universiteit Amsterdam

Biotechnology and the environment: research needs / edited by Gilbert S. Omenn and Albert H. Teich. Published/Created: Park Ridge, N.J.: Noyes Data Corp., 1986. Related Authors: Omenn, Gilbert S. Teich, Albert H. Description: x, 169 p.: ill.; 25 cm. ISBN: 0815511051: Notes: Includes bibliographies. Subjects: Biotechnology--Environmental aspects--Congresses. LC Classification: TP248.13 .B57 1986 Dewey Class No.: 363.7/01 19

Biotechnology and the environment: risk & regulation: proceedings of a seminar series / conducted by the American Association for the Advancement of Science for the U.S. Environmental Protection Agency; edited by Albert H. Teich, Morris A. Levin, Jill H. Pace. Published/Created: Washington, D.C.: AAAS, c1985. Related Authors: Teich, Albert H. Levin, Morris A. Pace, Jill H. American Association for the Advancement of Science. United States. Environmental Protection Agency. Description: xi, 201 p.: ill.; 23 cm.

ISBN: 0871682796 (pbk.) Notes: Seminars held in Washington, D.C., Nov. 1982-Oct. 1983. Includes bibliographies. Subjects: Biotechnology--Congresses. Biotechnology--Environmental aspects--Congresses. Biotechnology--Governmental policy--United States Congresses. Environmental policy--United States--Congresses. LC Classification: TP248.2 .B5513 1985 Dewey Class No.: 363.7/01 19

Biotechnology applications of microinjection, microscopic imaging, and fluorescence / edited by Peter H. Bach ... [et al.]. Published/Created: New York: Plenum Press, c1993. Related Authors: Bach, P. H. (Peter H.) European Workshop on Microscopic Imaging, Fluorescence, and Microinjection in Biotechnology (1st: 1992: London, England) Description: xi, 255 p.: ill.; 26 cm. ISBN: 0306444976 Notes: "Proceedings of the First European Workshop on Microscopic Imaging, Fluorescence, and Microinjection in Biotechnology, held April 21-24, 1992, in London, United Kingdom"--T.p. verso. Includes bibliographical references and index. Subjects: Microinjections--Congresses. Confocal microscopy--Congresses. Fluorescence spectroscopy--Congresses. Molecular probes--Congresses. Biotechnology--Methodology--Congresses. LC Classification: TP248.24 .B563 1993 Dewey Class No.: 610/.72 20

Biotechnology in tall fescue improvement / editor, M.J. Kasperbauer. Published/Created: Boca Raton: CRC Press, c1990. Related Authors: Kasperbauer, M. J., 1929- Description: 199 p.: ill.; 24 cm. ISBN: 0849348919

Contents: Importance and problems of tall fescue / Robert F. Barnes -- Potential biotechnological approaches / Karan Kaul -- Somatic cell culture and embryogenesis in the Poaceae / D.J. Gray -- Plant regeneration and evaluation / M.J. Kasperbauer -- Haploids / M.J. Kasperbauer -- Doubled haploids / M.J. Kasperbauer -- Cytology of haploids and doubled haploids / Warren D. Springer / Physiology and biochemistry / C.J. Nelson and D.A. Sleper -- Breeding and genetics / D.A. Sleper and C.J. Nelson. Notes: Includes bibliographical references and index. Subjects: Tall fescue--Biotechnology. LC Classification: SB201.T34 B56 1990 Dewey Class No.: 633.2/8 20

Biotechnology in the marine sciences: proceedings of the first annual MIT Sea Grant lecture and seminar / edited by Rita R. Colwell, Anthony J. Sinskey, E. Ray Pariser. Published/Created: New York: Wiley, c1984. Related Authors: Colwell, Rita R., 1934- Sinskey, Anthony J. Pariser, Ernst R. Massachusetts Institute of Technology. Sea Grant College Program. Description: xvii, 293 p.: ill.; 24 cm. ISBN: 0471882763 Notes: Held Mar. 18-20, 1982. "A Wiley-Interscience publication." Includes bibliographies and index. Subjects: Biotechnology--Congresses. Aquaculture--Congresses. Marine fouling organisms--Congresses. Marine pharmacology--Congresses. Series: Advances in marine science and biotechnology LC Classification: TP248.2 .B56 1984 Dewey Class No.: 660/.6/09162 19

Biotechnology of waste treatment and exploitation / [edited by] J.M. Sidwick and R.S. Holdom. Published/Created: Chichester, West Sussex, England: E.

Horwood; New York: Halsted Press, 1987. Related Authors: Sidwick, J. M. (John M.) Holdom, R. S. (Roger S.) Description: 332 p.: ill.; 25 cm. ISBN: 0853129177: 0470210311 (Halsted Press) Notes: Includes bibliographies and index. Subjects: Sewage-- Purification--Biological treatment. Biotechnology. Biochemistry. Series: Ellis Horwood series, water and wastewater technology. Variant Series: Ellis Horwood series in water and wastewater technology LC Classification: TD755 .B485 1987 Dewey Class No.: 628.3/51 19

Biotechnology: a comprehensive treatise in 8 volumes / edited by H.-J. Rehm and G. Reed. Published/Created: Weinheim [Germany]; Deerfield Beach, Fla.: Verlag Chemie, c1981- Related Authors: Rehm, Hans-Jürgen. Reed, Gerald, 1913- Description: v. <1-5, 6a-b, 7a, 8; in 9: ill.; 25 cm. ISBN: 0895730413 (U.S.: set) Incomplete Contents: v. 1. Microbial fundamentals -- v. 2. Fundamentals of biochemical engineering / volume editor, H. Brauer -- v. 3. Biomass, microorganisms for special applications, microbial products I, energy from renewable resources -- v. 4. Microbial products II / volume editors, H. Pape and H.-J. Rehm -- v. 5. Food and feed production with microorganisms -- v. 6a. Biotransformations / editor, K. Kieslich -- v. 6b. Special microbial processes / editor, Hans-Jürgen Rehm. -- v. 7a. Enzyme technology / volume editor, John F. Kennedy -- v. 8. Microbial degradations / volume editor, W. Schönborn. Notes: Includes bibliographies and indexes. Subjects: Biotechnology. LC Classification: TP248.2 .B5 1981 Dewey Class No.:

660/.62 19

Biotechnology: a multi-volume comprehensive treatise / edited by H.-J. Rehm and G. Reed, in cooperation with A. Pühler and P. Stadler. Edition Information: 2nd, completely rev. ed. Published/Created: Weinheim [Germany]; New York: VCH, c1991- Related Authors: Rehm, Hans-Jürgen. Reed, Gerald, 1913- Description: v. <1-5 pt. A-B, 7-8 pt. B, 10, lla-c, 13: ill.; 25 cm. ISBN: 3527283110 (v. 1: Weinheim: alk. paper) 156081151X (v. 1: N.Y.: alk. paper) 3527283129 (v. 2: Weinheim: alk. paper) 1560811528 (v. 2: N.Y.: alk. paper) 3527283137 (v. 3: Weinheim: alk. paper) 1560811536 (v. 3: N.Y.: alk. paper) 3527283285 (v. 5: Weinheim: alk. paper) 352728317X (v. 7: Weinheim: alk. paper) Incomplete Contents: v. 1. Biological fundamentals / edited by H. Sahm -- v. 2. Genetic fundamentals and genetic engineering / edited by A. Pühler -- v. 3. Bioprocessing / edited by G. Stephanopoulos -- v. 4. Measuring, modelling, and control / edited by K. Schügerl -- v. 5a. Recombinant proteins, monoclonal antibodies, and therapeutic genes / edited by A. Mountain, U. Ney, and D. Schomburg --v. 5b. Genomics and bioinformatics / edited by C.W. Sensen -- v. 7. Products of secondary metabolism / edited by H. Kleinkauf and H. von Döhren -- v. 8a. Biotransformation I / edited by D.R. Kelly -- v. 8b. Biotransformation II / vol. editor, D.R. Kelly -- v. 10. Special processes / edited by H.-J. Rehm -- v. 11a. Environmental processes I: wastewater treatment / edited by J. Winter -- llb. Environmental processes II: soil decontamination / edited by J. Klein -- 11c. Environmental processes III: solid waste and waste gas treatment, preparation of drinking water / edited by J. Klein and J. Winter -- v. [13]. Cumulative index/ edited by H.-J. Rehm and G. reed in cooperation with A. Pühler and P. Stadler. Notes: Vol. 7 imprint: Weinheim: VCH. Vol. 5 imprint: Wiley-VCH. Includes bibliographical references and indexes. Subjects: Biotechnology. LC Classification: TP248.2 .B5465 1991 Dewey Class No.: 660/.6 20

Campbell, Ailsa M. Monoclonal antibody technology: the production and characterization of rodent and human hybridomas / Ailsa M. Campbell. Published/Created: Amsterdam; New York: Elsevier, 1984. Description: xiv, 265 p.: ill.; 21 cm. ISBN: 0444805923 0444805753 (pbk.) Notes: Includes index. Bibliography: p. 243-254. Subjects: Monoclonal antibodies. Hybridomas. Biotechnology. Antibodies, Monoclonal. Hybidomas. Immunologic Technics. Series: Laboratory techniques in biochemistry and molecular biology; v. 13 LC Classification: QP519 .L2 vol. 13 QR186.85 Dewey Class No.: 615/.37 19

Cellulose sources and exploitation: industrial utilization, biotechnology, and physico-chemical properties / editors, J.F. Kennedy, G.O. Phillips, P.A. Williams. Published/Created: New York: Ellis Horwood, 1990. Related Authors: Kennedy, John F., 1942- Phillips, Glyn O. Williams, Peter A. Description: xviii, 520 p.: ill.; 25 cm. ISBN: 0131219553: Subjects: Cellulose. Cellulose--Biotechnology. Series: Ellis Horwood series in polymer science and technology LC Classification: TP248.65.C45 C42 1990 Dewey Class No.: 661/.802 20

Cellulose: structural and functional aspects / editors, J.F. Kennedy, G.O. Phillips, P.A. Williams. Published/Created: Chichester: Ellis Horwood; New York: Halsted Press, 1989. Related Authors: Kennedy, John F., 1942- Phillips, Glyn O. Williams, Peter A. Cellucon Conferences (Organization) Sen'i Gakkai (Japan) Cellucon 88 Japan (1988: Kyoto, Japan) Related Titles: [Wood processing and utilization. Description: xi, 519 p.: ill.; 25 cm. ISBN: 0745804713: 0470216131 (Halsted Press) Notes: "This book and its companion volume, Wood processing and utilization, present the proceedings of the Cellucon 88 Conference whice was held Kyoto, Japan"--Cover p. [21]. "This Cellucon-Japan Conference was a joint venture between Cellucon Conferences and the Japan Society of Fibre Science and Technology"--P. 2. Includes bibliographical references and index. Subjects: Cellulose. Cellulose--Biotechnology. Series: Ellis Horwood series in polymer science and technology LC Classification: TP248.65.C45 C43 1989 Dewey Class No.: 661/.802482 20

Chemical engineering problems in biotechnology / edited by M.A. Winkler. Published/Created: London; New York: Published for the Society of Chemical Industry by Elsevier Applied Science, c1990. Related Authors: Winkler, M. A. (Michael A.), 1935- Description: viii, 370 p.: ill.; 23 cm. ISBN: 1851664548 Notes: Includes bibliographical references (p. 337-350) and index. Subjects: Biochemical engineering. Biotechnology. Series: Critical reports on applied chemistry; v. 29 LC Classification: TP248.3 .C47

1990 Dewey Class No.: 660/.63 20

Chemicals from microalgae / edited by Zvi
Cohen. Published/Created: London;
Philadelphia, PA: Taylor & Francis,
c1999. Related Authors: Cohen, Zvi,
Ph. D. Description: xvii, 419 p.: ill.
(some col.); 26 cm. ISBN: 0748405151
Notes: Includes bibliographical
references and index. Subjects:
Microalgae--Biotechnology. Algae--
chemistry. Biological Factors--
biosynthesis. Biotechnology--methods.
Microalgae--Biotechnology. LC
Classification: TP248.27.A46 C48 1999
Dewey Class No.: 660.6 21

Chemistry and biotechnology of biologically
active natural products: proceedings of
the fourth international conference
Budapest, 10-14 August 1987 / editor,
Cs. Szántay; associate editors, Á.
Gottsegen, G. Kovács.
Published/Created: Budapest:
Akadémiai Kiadó, 1988. Related
Authors: Szántay, Csaba. Gottsegen, Á.
Kovács, G. Federation of European
Chemical Societies. International
Conference on Chemistry and
Biotechnology of Biologically Active
Natural Products (4th: 1987: Budapest,
Hungary) Description: xi, 402 p.; 25
cm. ISBN: 9630549506 Notes: Papers
from the Fourth International
Conference on Chemistry and
Biotechnology of Biologically Active
Natural Products, held under the
auspices of the Federation of European
Chemical Societies. Includes
bibliographies and index. Subjects:
Biotechnology--Congresses. Natural
products--Congresses. LC
Classification: TP248.14 .C48 1988
Dewey Class No.: 660/.6 20

Chemistry and biotechnology of biologically
active natural products: proceedings of
the second international conference,
Budapest, 15-19 August 1983 / edited
by Cs. Szántay; associate editors, Á.
Gottsegen and G. Kovács.
Published/Created: Amsterdam; New
York: Elsevier Science Publishers; New
York, N.Y., U.S.A.: Distribution for the
U.S.A. and Canada, Elsevier Science
Pub. Co., 1984. Related Authors:
Szántay, Csaba. Gottsegen, Á. Kovács,
G. Federation of European Chemical
Societies. International Conference on
Chemistry and Biotechnology of
Biologically Active Natural Products
(2nd: 1983: Budapest, Hungary)
Description: xvii, 377 p.; 25 cm. ISBN:
0444996087 (U.S.): Notes: Papers
presented at the Second International
Conference on Chemistry and
Biotechnology of Biologically Active
Natural Products, held Aug. 15-19, 1983
in Budapest, Hungary, sponsored by the
Federation of European Chemical
Societies. Includes bibliographical
references and index. Subjects:
Biotechnology--Congresses. Natural
products--Congresses. Series: Studies in
organic chemistry (Elsevier Science
Publishers); 17. Variant Series: Studies
in organic chemistry; 17 LC
Classification: TP248.2 .C49 1984
Dewey Class No.: 660/.6 19

Chinese journal of biotechnology.
Published/Created: New York: Allerton
Press, [c1989- Related Titles: [Sheng
wu kung ch`eng hsüeh pao. English.
Description: v.: ill.; 26 cm. Carries the
numbering of the journal from which
selections are taken; issues for v. 4-
called also: 1988- . Vol. 4, no. 1-
Current Frequency: 4 no. a year, <1990-
Former Frequency: Semiannual, <1989-
ISSN: 1042-749X Cancel/Invalid

LCCN: sn 89006357 Notes: SERBIB/SERLOC merged record Translation of selected articles from: Sheng wu kung ch`eng hsüeh pao. Indexed entirely by: Index medicus 0019-3879 1988- Subjects: Biotechnology--Periodicals. Biotechnology--China--Periodicals. Biotechnology--periodicals. Genetic Techniques--periodicals. LC Classification: TP248.13 .C48 Dewey Class No.: 660/.6 20

Current biotechnology abstracts. Published/Created: Nottingham, England: Royal Society of Chemistry, c1983-c1990. Description: 8 v.; 30 cm. Issue 1 (Apr. 1983)-v. 8, issue 12 (Dec. 1990). Current Frequency: Monthly, 1984-1990 Former Frequency: Nine no. a year, Apr. 1983-Dec. 1983 Continued by: Current biotechnology (DLC) 93660715 (OCoLC)23854566 ISSN: 0264-3391 CODEN: CBABD3 Notes: Title from cover. SERBIB/SERLOC merged record Subjects: Biotechnology--Abstracts--Periodicals. Biotechnology--indexes. LC Classification: TP248.2 .C87 Dewey Class No.: 660.6/05 19

Current biotechnology. Published/Created: Cambridge, UK: Royal Society of Chemistry, c1991- Related Authors: Royal Society of Chemistry (Great Britain) Description: v.; 30 cm. Vol. 9, issue 1 (Jan. 1991)- Current Frequency: Monthly Continues: Current biotechnology abstracts 0264-3391 (DLC) 84641644 (OCoLC)10482375 ISSN: 0960-5037 Cancel/Invalid LCCN: sn 91033293 CODEN: CUBIER Notes: Title from cover. Produced from the Chemical engineering and biotechnology abstracts (CEBA) database. SERBIB/SERLOC merged record Subjects: Biotechnology--Abstracts--Periodicals. Biotechnology--abstracts. LC Classification: TP248.2 .C87 Dewey Class No.: 660.6/05 19

Cyanobacterial and algal metabolism and environmental biotechnology / editor, Tasneem Fatma. Published/Created: New Delhi: Narosa Pub. House, c1999. Related Authors: Fatma, Tasneem. Description: xi, 272 p.: ill.; 23 cm. ISBN: 8173192731 Summary: Contributed articles. Notes: Includes bibliographical references. Subjects: Cyanobacteria. Cyanobacteria--Biotechnology. Algae. Environmental biotechnology. LC Classification: QR99.63 .C93 1999 Dewey Class No.: 572/.42939 21

Derwent biotechnology abstracts. Published/Created: [London: Derwent Publications], 1982- Description: v.: ill.; 27 cm. Vol. 1, no. 1 (15 July 1982)- Current Frequency: Biweekly ISSN: 0262-5318 Cancel/Invalid LCCN: sn 84010673 Notes: SERBIB/SERLOC merged record Additional Form Avail.: Issued also on CD-ROM as: Derwent biotechnology abstracts database. Subjects: Biotechnology--Abstracts--Periodicals. Genetic engineering--Abstracts--Periodicals. Genetic recombination--Abstracts--Periodicals. Bioengineering--Abstracts--Periodicals. Fermentation--Abstracts--Periodicals. Biotechnology--abstracts. LC Classification: TP248.13 .D47 Dewey Class No.: 660.6/05 19

Doelle, H. W. Microbial process development / Horst W. Doelle. Published/Created: Singapore; River Edge, NJ: World Scientific, c1994. Description: xiii, 308 p.: ill.; 23 cm. ISBN: 9810215150 Notes: Includes bibliographical references (p. 293-300) and index. Subjects: Industrial microbiology. Microbial biotechnology. LC Classification: TP248.27.M53 D64 1994 Dewey Class No.: 660/.62 20

Edwards, M. J. (Malcolm John) ATCC microbes & cells at work: an index to ATCC strains with special applications / compiled from the ATCC catalogues by M.J. Edwards. Edition Information: 2nd ed. Published/Created: Rockville, Md.: American Type Culture Collection, c1991. Related Authors: American Type Culture Collection. Related Titles: [ATCC microbes & cells at work. ATCC microbes and cells at work. Microbes & cells at work. Description: vi, 305 p.; 28 cm. ISBN: 0930009401: Notes: Cover ATCC microbes and cells at work. Rev. ed. of: ATCC microbes & cells at work / edited by M.J. Edwards ... [et al.]. 1st ed. 1988. Includes bibliographical references (p. 170-305). Subjects: American Type Culture Collection--Catalogs. Biomolecules--Synthesis--Bibliography--Indexes. Microorganisms--Biotechnology--Catalogs and collections. Animal cell biotechnology--Catalogs and collections. LC Classification: Z7914.B32 A86 1991 TP248.2 Dewey Class No.: 660/.6/0294 20

Elkington, John. The gene factory: inside the genetic and biotechnology business revolution / John Elkington. Edition Information: 1st Carroll & Graf ed. Published/Created: New York: Carroll & Graf Publishers, 1985. Description: 240 p.; 23 cm. ISBN: 0881842087: Notes: Includes index. Bibliography: p. [232]-234. Subjects: Biotechnology industries. LC Classification: HD9999.B442 E42 1985 Dewey Class No.: 338.4/766606 19

Endress, R. (Rudolf), 1941- Plant cell biotechnology / R. Endress. Published/Created: Berlin; New York: Springer-Verlag, c1994. Description: xiv, 353 p.: ill.; 24 cm. ISBN: 3540569472 (Berlin: alk. paper) 0387569472 (New York: alk. paper) Notes: Includes bibliographical references and index. Subjects: Plant biotechnology. LC Classification: TP248.27.P55 E53 1994 Dewey Class No.: 660/.6 20

Enei, Hitoshi. Recent progress in microbial production of amino acids / by Hitoshi Enei, Kenzo Yokozeki, Kunihiko Akashi. Published/Created: New York: Gordon and Breach, c1989. Related Authors: Yokozeki, Kenzo. Akashi, Kunihiko. Description: xiii, 140 p.: ill.; 22 cm. ISBN: 288124324X Notes: Includes bibliographical references. Subjects: Amino acids--Biotechnology. Amino acids--Synthesis. Microbiological synthesis. Series: Japanese technology reviews; vol. 5 LC Classification: TP248.65.A43 E54 1989

Engineering of/with lipases / edited by F. Xavier Malcata. Published/Created: Dordrecht; Boston: Kluwer Academic, c1996. Related Authors: Malcata, F. Xavier. North Atlantic Treaty Organization. Scientific Affairs Division. NATO Advanced Study Institute on Engineering of/with Lipases (1995: Póvoa de Varzim, Portugal) Description: xv, 618 p.: ill.; 25 cm. ISBN: 0792340035 (hb: alk. paper) Notes: "Published in cooperation with NATO Scientific Affairs Division." "Proceedings of the NATO Advanced Study Institute on Engineering of/with Lipases, Póvoa de Varzim, Portugal, May 22-June 2, 1995"--T.p. verso. Includes bibliographical references.

Subjects: Lipases--Biotechnology.
Series: NATO ASI series. Series E,
Applied sciences; no. 317. Variant
Series: NATO ASI series. Series E,
Applied sciences; vol. 317 LC
Classification: TP248.65.E59 E54 1996
Dewey Class No.: 660/.634 20

Environment and biotechnology: a new
frontiers of plant pathology / editors,
B.P. Singh, H.N. Verma, K.M.
Srivastava. Published/Created: New
Delhi: Today & Tomorrow's Printers &
Publishers; Houston, Tex.: Distributed
in USA by Scholarly Publications,
1988. Related Authors: Singh, B. P.
Verma, H. N. Srivastava, K. M. Group
Discussion Meeting on "Environmental
Pathology and Role of Biotechnology in
Plant Pathology" (1984: National
Botanical Research Institute)
Description: 126 p.: ill.; 26 cm. ISBN:
8170193230: 1555281516 (U.S.) Notes:
Title corrected by label. Papers
presented at the Group Discussion
Meeting on "Environmental Pathology
and Role of Biotechnology in Plant
Pathology", held at the National
Botanical Research Institute, 1984.
Includes bibliographies. Subjects: Plant
diseases--Congresses. Plant diseases--
Environmental aspects--Congresses.
Phytopathogenic microorganisms--
Ecology--Congresses. Agricultural
biotechnology--Congresses. Crops--
Effect of air pollution on--Congresses.
Series: Recent researches in ecology,
environment, and pollution; 2 LC
Classification: SB727 .E58 1988 Dewey
Class No.: 632 20

Environmental biotechnology and cleaner
bioprocesses / edited by Eugenia J.
Olguin, Gloria Sanchez, Elizabeth
Hernandez. Published/Created: New
York: Taylor & Francis, c2000. Related

Authors: Olguín, Eugenia. Sanchez,
Gloria. Hernandez, Elizabeth.
Description: xx, 319 p.: ill.; 26 cm.
ISBN: 0748407294 (alk. paper) Notes:
Includes bibliographical references and
index. Subjects: Bioremediation.
Biotechnology. Production
management--Environmental aspects.
Sustainable development. LC
Classification: TD192 .E584 2000
Dewey Class No.: 628.5 21

Environmental effects of transgenic plants:
the scope of adequacy and regulation /
Committee on Environmental Impacts
associated with Commercialization of
Transgenic Plants Board on Agriculture
and Natural Resources Division on
Earth and Life Studies, National
Research Council. Published/Created:
Washington, D.C.: National Academy
Press, 2002. Projected Pub. Date: 1111
Related Authors: National Research
Council. Committee on Environmental
Impacts. Description: p. cm. ISBN:
0309082633 (hardcover) Notes:
Includes bibliographical references (p.).
Subjects: Transgenic plants--Risk
assessment. Agricultural biotechnology-
-Environmental aspects. LC
Classification: SB123.57 .E58 2002
Dewey Class No.: 631.5/233 21

Environmental impacts of aquatic
biotechnology. Published/Created:
Paris, France: Organisation for
Economic Co-operation and
Development; Washington, D.C.:
OECD Publications and Information
Center [distributor], c1995. Related
Authors: Organisation for Economic
Co-operation and Development. OECD
Workshop on Environmental Impacts of
Aquaculture Using Aquatic Organisms
Derived through Modern Biotechnology
(1993: Trondheim, Norway)

Description: 171 p.: ill.; 27 cm. ISBN: 9264146660 Contents: The present state of sea ranching of fish and shellfish in Japan / Hisashi Kanno -- Genetically modified fish populations / D. J. Penman, M. Woodwark and B. J. McAndrew -- Diversity of microalgae and their possible application / Shigetoh Miyachi -- Gene flow and analysis of structured populations / Chris Gliddon and Jérôme Goudet -- Implications of introduction of transgenic fish into natural ecosystems / Anne R. Kapuscinski -- Modern biotechnology and its application to macroalgae cultivation in Japan / Masahiro Notoya -- Possible genetic and ecological effects of escaped salmonids in aquaculture / Øystein Skaala -- Modern biotechnology and its application to aquaculture / Tadashi Matsunaga -- The use of microalgae and microorganisms in rearing systems for fish larvae and juveniles / Osamu Imada -- Triploidy in the Manila Clam (tapes philippinarum) / Susan Utting -- Modern biotechnology and its application to shellfish / Katsuhiko T. Wada -- Sex control and ploidy manipulation in sea bass / Manuel Carrillo, S. Zanuy, M. Blázquez, J. Ramos, F. Piferrer and E. M. Donaldson. Notes: Papers from the OECD Workshop on Environmental Impacts of Aquaculture Using Aquatic Organisms Derived through Modern Biogechnology, held in Trondheim, Norway, 9-11 June 1993. Includes bibliographical references. Subjects: Marine biotechnology--Environmental aspects--Congresses. Biotechnology--Environmental aspects--Congresses. Aquatic biology--Congresses. Series: OECD documents LC Classification: TP248.27.M37 E56 1995

Environmental monitoring and biodiagnostics of hazardous contaminants / edited by Michael Healy, Donald L. Wise, and Murray Moo-Young. Published/Created: Boston: Kluwer Academic Publishers, 2001. Projected Pub. Date: 01 Related Authors: Healy, Michael, 1950- Wise, Donald L. (Donald Lee), 1937- Moo-Young, Murray. Description: p. cm. ISBN: 079236869X (alk. paper) Subjects: Industrial microbiology. Hazardous wastes--Biodegradation. Microbial ecology. Microbial biotechnology. LC Classification: QR53 .E585 2001 Dewey Class No.: 660.6/2 21

Enzymatic degradation of insoluble carbohydrates / John N. Saddler, editor, Michael H. Penner, editor. Published/Created: Washington, DC: American Chemical Society, 1995. Related Authors: Saddler, John N., 1953- Penner, Michael Henry. American Chemical Society. Division of Agricultural and Food Chemistry. American Chemical Society. Meeting (207th: 1994: San Diego, Calif.) Description: ix, 374 p.: ill.; 24 cm. ISBN: 0841233411 Notes: "Developed from a symposium sponsored by the Division of Agricultural and Food Chemistry at the 207th National Meeting of the American Chemical Society, San Diego, California, March 13-17, 1994. Includes bibliographical references and indexes. Subjects: Polysaccharides--Biotechnology--Congresses. Polysaccharides--Metabolism--Congresses. Glucosidases--Biotechnology--Congresses. Cellulase--Biotechnology--Congresses. Series: ACS symposium series; 618 LC Classification: TP248.65.P64 E59 1995

Dewey Class No.: 661.8/02 20

Enzyme and microbial technology.
Published/Created: Guildford, Eng. IPC
Science and Technology Press.
Description: v. ill. 30 cm. v. 1- Jan.
1979- Current Frequency: Monthly,
1984- Former Frequency: Bimonthly,
1982-1983 Quarterly, 1979-1981 ISSN:
0141-0229 Cancel/Invalid LCCN: sn
79000879 CODEN: EMTED2 Notes:
SERBIB/SERLOC merged record
Indx'd selectively by: Biological
abstracts 0006-3169 Chemical abstracts
0009-2258 Energy information abstracts
0147-6521 Engineering index annual
(1968) 0360-8557 Engineering index
monthly (1984) 0742-1974 Engineering
index bioengineering abstracts 0736-
6213 Engineering index energy
abstracts 0093-8408 Environment
abstracts 0093-3287 Excerpta medica
BIOTECHSEEK 1990- Life sciences
collection Reference sources 0163-3546
Additional Form Avail.: Available also
by subscription via the World Wide
Web. Subjects: Enzymes--Industrial
applications--Periodicals.
Biotechnology--Periodicals.
Microbiological Techniques--
periodicals. Microbiology--periodicals.
Technology--periodicals. LC
Classification: TP248.E5 E565 Dewey
Class No.: 660/.6/05

Enzyme applications in fiber processing /
Karl-Erik L. Eriksson, editor, Artur
Cavaco-Paulo, editor.
Published/Created: Washington, DC:
American Chemical Society, c1998.
Related Authors: Eriksson, Karl-Erik.
Cavaco-Paulo, Artur. ACS Symposium
Enzyme Applications in Fiber
Processing (1997: San Francisco, Calif.)
Description: xii, 340 p.: ill.; 24 cm.
ISBN: 0841235473 (acid-free paper)

Notes: "This book presents papers from
the ACS symposium "Enzyme
Applications in Fiber Processing" held
in San Francisco, April 1997"--Pref.
Includes bibliographical references and
index. Subjects: Wood-pulp--
Biotechnology--Congresses. Enzymes--
Biotechnology--Congresses. Textile
fibers--Congresses. Microbial enzymes-
-Industrial applications--Congresses.
Series: ACS symposium series; 687 LC
Classification: TS1176.6.B56 E58 1998
Dewey Class No.: 676/.12 21

Enzyme biotechnology: protein engineering,
structure prediction, and fermentation /
editor, M. James C. Crabbe.
Published/Created: New York: E.
Horwood, 1990. Related Authors:
Crabbe, M. James C. Description: 128
p., [4] p. of plates: ill. (some col.); 25
cm. ISBN: 0132823934 Notes: Includes
bibliographical references and index.
Subjects: Enzymes--Biotechnology.
Protein engineering. Series: Ellis
Horwood series in biochemistry and
biotechnology LC Classification:
TP248.65.E59 E59 1990 Dewey Class
No.: 660/.63 19

Enzyme engineering XIV / edited by Allen
I. Laskin, Gao-Xiang Li, and Yao-Ting
Yu. Published/Created: New York,
N.Y.: New York Academy of Sciences,
1998. Related Authors: Laskin, Allen I.,
1928- Li, Gao-xiang. Yu, Yao-ting.
International Enzyme Engineering
Conference (14th: 1997: Beijing, China)
Description: xviii, 665 p.: ill.; 24 cm.
ISBN: 1573311499 (cloth: alk. paper)
1573311502 (pbk.: alk. paper) Notes:
"This volume is the result of the
Fourteenth International Enzyme
Engineering Conference, which was
held on October 12-17, 1997, in Beijing,
China ..."--Contents p. Includes

bibliographical references and index. Subjects: Enzymes--Biotechnology--Congresses. Series: Annals of the New York Academy of Sciences; v. 864 LC Classification: Q11 .N5 vol. 864 TP248.65.E59 Dewey Class No.: 500 s 660.6/34 21

Enzyme engineering: immobilized biosystems / editor, Peter Gemeiner; translation editor, John L. Francis. Published/Created: New York: E. Horwood, 1992. Related Authors: Gemeiner, Peter, 1946. Description: 298 p.: ill.; 25 cm. ISBN: 0132782278 Notes: Translated from Czechoslovakian. Includes bibliographical references and index. Subjects: Immobilized enzymes--Biotechnology. Enzymes--Biotechnology. Biotechnology. Enzymes. Series: Ellis Horwood series in biochemistry and biotechnology LC Classification: TP248.65.I45 E59 1992 Dewey Class No.: 660/.634 20

Enzyme systems for lignocellulose degradation / edited by M.P. Coughlan. Published/Created: London; New York: Elsevier, c1989. Related Authors: Coughlan, Michael P. Description: x, 408 p.: ill.; 25 cm. ISBN: 1851664114 Notes: Includes bibliographical references. Subjects: Cellulase--Biotechnology. LC Classification: TP248.65.C44 E59 1989 Dewey Class No.: 661/.802 20

Enzymes for pulp and paper processing / Thomas W. Jeffries, Liisa Viikari, [editors]. Published/Created: Washington, DC: American Chemical Society, c1996. Related Authors: Jeffries, T. W. (Thomas William), 1947- Viikari, Liisa, 1949- American Chemical Society. Cellulose, Paper, and Textile Division. American Chemical Society. Meeting (211th: 1996: New Orleans, La.) Description: x, 326 p.: ill.; 24 cm. ISBN: 0841234787 (acid-free paper) Notes: Developed from a symposium sponsored by the Cellulose, Paper, and Textile Division at the 211th National Meeting of the American Chemical Society, New Orleans, Louisiana, March 24-28, 1996. Includes bibliographical references and indexes. Subjects: Wood-pulp--Biotechnology. Enzymes--Industrial applications. Papermaking--Chemistry. Series: ACS symposium series, 0097-6156; 655 LC Classification: TS1176.6.B56 E59 1996 Dewey Class No.: 676 20

Enzymes in action: green solutions for chemical problems / edited by Binne Zwanenburg, Marian Mikolajczyk, and Piotr Kielbasinski. Published/Created: Boston: Kluwer Academic Publishers, 2000. Related Authors: Zwanenburg, B. Mikolajczyk, Marian, 1937- Kielbasi´nski, Piotr. Description: ix, 458 p.: ill.; 25 cm. ISBN: 0792366956 (alk. paper) Notes: "Proceedings of the NATO Advanced Institute on Enzymes in Heteroatom Chemistry (Green Solutions for Chemical Problems), Berg en Dal, The Netherlands, 19-30 June 1999"--Verso t.p. Includes bibliographical references. Subjects: Enzymes--Biotechnology Organic compounds--Synthesis. Organic compounds--Biodegradation. Chemical agents (Munitions)--Biodegradation. Series: NATO ASI series. Partnership sub-series 1, Disarmament technologies; vol. 33 LC Classification: TP248.65.E59 E59185 2000 Dewey Class No.: 660.6/34 21

Enzymes in carbohydrate synthesis / Mark D. Bednarski, editor, Ethan S. Simon, editor. Published/Created: Washington, DC: American Chemical Society, 1991. Related Authors: Bednarski, Mark D., 1958- Simon, Ethan S., 1963- American Chemical Society. Division of Carbohydrate Chemistry. American Chemical Society. Meeting (199th: 1990: Boston, Mass.) Description: xi, 131 p.: ill.; 24 cm. ISBN: 0841220972 (alk. paper) Notes: "Developed from a symposium sponsored by the Division of Carbohydrate Chemistry at the 199th National Meeting of the American Chemical Society, Boston, Massachusetts, April 22-27, 1990." Includes bibliographical references and indexes. Subjects: Enzymes--Biotechnology--Congresses. Carbohydrates--Biotechnology--Congresses. Carbohydrates--Synthesis--Congresses. Series: ACS symposium series, 0097-6156; 466 LC Classification: TP248.65.E59 E592 1991 Dewey Class No.: 660/.634 20

Enzymes in industry: production and applications / edited by Wolfgang Gerhartz. Published/Created: Weinheim, F.R.D.; New York, NY, USA: VCH, c1990. Related Authors: Gerhartz, Wolfgang. Description: xvii, 321 p.: ill.; 25 cm. ISBN: 0895739372 (U.S.: alk. paper) Notes: Includes bibliographical references (p. [255-298) and index. Subjects: Enzymes--Biotechnology. Enzymes--Industrial applications. LC Classification: TP248.65.E59 E593 1990 Dewey Class No.: 660/.634 20

Enzymes in lipid modification / edited by Uwe T. Bornscheuer in collaboration with the German Society for Fat Science (DGF). Published/Created: Weinheim; New York: Wiley-VCH, c2000. Related Authors: Bornscheuer, U. T. (Uwe Theo), 1964- Deutsche Gesellschaft für Fettwissenschaft. Description: xx, 424 p.: ill.; 25 cm. ISBN: 3527301763 (alk. paper) Contents: Lipases. The exploitation of lipase selectivities for the production of acylglycerols / Rob Diks and John Bosley -- Fractionation of fatty acids and other lipids using lipases / Kumar D. Mukherjee -- Lipid modification in water-in-oil microemulsions / Douglas G. Hayes -- Cloning, mutagensis and biochemical properties of a lipase from the fungus rhizopus delemar / Michael J. Haas...[et al.] -- Molecular basis of specificity and stereoselectivity of microbial lipases toward triacylglycerols / Jürgen Pleiss -- Lipase-catalyzed synthesis of regioisomerically pure mono- and diglycerides / B. Aha...[et al.] -- Lipase-catalyzed peroxy fatty acids generation in lipid oxidation / M. Rüsch gen. Klaas and Siegfried Warwel -- Production of functional lipids containing polyunsaturated fatty acids with lipase / Yuji Shimada, Akio Sugihara and Yoshio Tominaga -- Lipase-catalyzed synthesis of structured triacylglycerols containing polyunsaturated fatty acids-- monitoring of the reaction and increasing the yield / Tsuneo Yamane -- Enrichment of lipids with EPA and DHA by lipase / Gudmundur G. Haraldsson -- Modification of oils and fats by lipase-catalyzed interesterification: aspects of process engineering / Xuebing Xu -- Phospholipases. Phospholipases used in lipid transformations / Renate Ulbrich-Hofmann -- Preparation and application of immobilized phospholipases / Peter Grunwald -- Enzymatic conversions of glycerophospholipids / Patrick Adlercreutz -- Lipoxygenases. Application of lipoxygenases and related enzymes for the preparation of oxygenated lipids / Ivo Feussner and Hartmut Kuehn / Properties and applications of lipoxygenases / Gilles Iacazio and Dominique Martini-Iacazio -- Miscellaneous enzymes. Enzymatic synthesis and modification of glycolopids / Siegmund Lang, Christoph Syldatk and Udo Rau -- Fatty acid hydroxylations using P450 monooxygenases / Ulrich Schwaneberg and Uwe T. Bornscheuer. Notes: Includes bibliographical references and index. Subjects: Lipids--Metabolism. Lipids--Biotechnology. Enzymes-- Biotechnology. Lipids--Metabolism. Lipids--Biotechnology. Enzymes-- Biotechnology. LC Classification: QP751 .E586 2000 Dewey Class No.: 572/.574 21

Federation of Asian and Oceanian Biochemists. Symposium. (7th: 1988: Kuala Lumpur, Malaysia) Advances in biochemistry and biotechnology in Asia and oceania: proceedings of the 7th Federation of Asian and Oceanian Biochemists Symposium, 28-30 November 1988 held at Kuala Lumpur, Malaysia / edited by A. Sipat ...[et al.]. Published/Created: [Kuala Lumpur]: FAOB, [1988?] Related Authors: A. Sipat (Abdullah Sipat) Description: 1 v. (various pagings): ill.; 23 cm. Notes: Includes bibliographical references. Subjects: Biochemistry--Congresses. Biotechnology--Congresses. LC Classification: QP501 .F46 1988 Dewey Class No.: 574.19/2 20

Federation of European Microbiological Societies. Symposium (1988: Troia, Portugal) Microbiology of extreme environments and its potential for biotechnology: proceedings of the Federation of European Microbiological Societies Symposium held in Troia, Portugal, 18-23 September 1988 / edited by M.S. da Costa, J.C. Duarte, and R.A.D. Williams. Published/Created: London; New York: Elsevier Applied Science, c1989. Related Authors: Costa, M. S. da. Duarte, J. M. Cardoso (José M. Cardoso) Williams, R. A. D. (Ralph Anthony David), 1938- Description: xiv, 429 p.: ill.; 23 cm. ISBN: 1851663614 Notes: Includes bibliographies. Subjects: Microbial ecology--Congresses. Microbial biotechnology--Congresses. Series: FEMS symposium; no. 49 LC Classification: QR100 .F43 1988 Dewey Class No.: 576/.15 20

Filamentous fungi / edited by D.L. Hawksworth and B.E. Kirsop, in collaboration with S.C. Jong ... [et al.]. Published/Created: Cambridge, [England]; New York: Cambridge University Press, 1988. Related Authors: Hawksworth, D. L. Kirsop, B. E. Jong, S. C. Description: xii, 209 p.: ill.; 24 cm. ISBN: 0521352266 Notes: Includes bibliographies and index. Subjects: Filamentous fungi-- Biotechnology. Series: Living resources for biotechnology LC Classification: TP248.27.F86 F55 1988 Dewey Class No.: 660/.62 19

Fitch, J. Patrick. An engineering introduction to biotechnology / by J. Patrick Fitch. Published/Created: Bellingham, WA: SPIE Optical Engineering Press, 2002. Projected Pub. Date: 0201 Description: p. cm. ISBN: 0819444979 Notes: Includes bibliographical references and index. Subjects: Biotechnology. Genetic engineering. Series: SPIE tutorial texts series; TT55 LC Classification: TP248.2 .F55 2002 Dewey Class No.: 660.6 21

Foreign gene expression in fission yeast: Schizosaccharomyces pombe / Yuko Giga-Hama, Hiromichi Kumagai, (eds.). Published/Created: Berlin; New York: Springer; Georgetown, TX: Landes Bioscience, c1997. Related Authors: Giga-Hama, Yuko, 1959- Kumagai, Hiromichi, 1954- Description: 182 p.: ill. (some col.); 24 cm. ISBN: 3540632700 (alk. paper) Notes: Includes bibliographical references and index. Subjects: Schizosaccharomyces pombe. Yeast fungi--Biotechnology. Yeast fungi--Genetic engineering. Recombinant proteins. Series: Biotechnology intelligence unit (Unnumbered) Variant Series: Biotechnology intelligence unit LC Classification: TP248.27.Y43 F67 1997 Dewey Class No.: 660/.65 21

From clone to clinic / edited by D.J.A. Crommelin and H. Schellekens. Published/Created: Dordrecht, The Netherlands; Boston: Kluwer Academic Publishers, c1990. Related Authors: Crommelin, D. J. A. (Daan J. A.) Schelleken, Huub. Description: xvi, 378 p.: ill.; 25 cm. ISBN: 0792309456 (alk. paper) Notes: Selection of papers presented at a meeting held in Amsterdam in Mar. 1990. Includes bibliographical references. Subjects: Pharmaceutical biotechnology. Monoclonal antibodies. Antibodies, Monoclonal--therapeutic use. Biological Products--standards. Cloning, Molecular. DNA, Recombinant. Genetic Engineering. Series: Developments in biotherapy; v. 1 LC Classification: RS380 .F76 1990 Dewey Class No.: 615/.7 20

From ethnomycology to fungal biotechnology: exploiting fungi from natural resources for novel products / edited by Jagjit Singh and K.R. Aneja. Published/Created: New York: Kluwer Academic/Plenum Publishers, c1999. Related Authors: Singh, Jagjit, 1956- Aneja, K. R. Description: xiii, 293 p.: ill.; 26 cm. ISBN: 0306460599 Notes: "Proceedings of the International Conference 'From Ethnomycology to Fungal Biotechnology: Exploiting Fungi from Natural Resources for Novel Products', held December 15-16, 1997, in Simla, India"--T.p. verso. Includes bibliographical references and index. Subjects: Mycology--Congresses. Fungi--Biotechnology--Congresses. Mycorrhizal fungi--Congresses. Fungi as biological pest control agents--Congresses. Fungi--Therapeutic use--Congresses. Phytopathogenic fungi--Congresses. LC Classification: QK600.3 .F76 1999 Dewey Class No.: 579.5 21

From gene to protein: translation into biotechnology / edited by Fazal Ahmad ... [et al.]. Published/Created: New York: Academic Press, 1982. Related Authors: Ahmad, Fazal, 1935- Description: xxii, 589 p.: ill.; 24 cm. ISBN: 0120455609 Notes: Proceedings of the symposium held in Miami Beach in Jan. 1982. Includes bibliographical references and index. Subjects: Genetic engineering--Congresses. Proteins--Biotechnology--Congresses. Series: Miami winter symposia; v. 19 LC Classification: QH442 .F75 1982 Dewey Class No.: 660/.6 19

From genetic experimentation to biotechnology--the critical transition: proceedings of a symposium, From genetic experimentation to biotechnology--the critical transition, held at Consiglio nazionale delle ricerche, Piazzale Aldo Moro, 7, Rome,

Italy, 20-23 September 1981 / edited by William J. Whelan, Sandra Black. Published/Created: Chichester [Sussex]; New York: Wiley, c1982. Related Authors: Whelan, W. J. (William Joseph) Black, Sandra. Consiglio nazionale delle ricerche (Italy) Committee on Genetic Experimentation. Description: xx, 266 p.: ill.; 24 cm. ISBN: 0471101486: Notes: "A Wiley-Interscience publication." An International symposium sponsored by the Consiglio nazionale delle ricerche, Rome and the Committee on Genetic Experimentation. Includes bibliographies and index. Subjects: Genetic engineering--Congresses. Recombinant DNA--Congresses. Biotechnology--Congresses. Genetic engineering--Industrial applications--Congresses. Recombinant DNA--Industrial applications--Congresses. Genetic engineering--Social aspects--Congresses. Recombinant DNA--Social aspects--Congresses. LC Classification: QH442 .F76 1982 Dewey Class No.: 660/.6 19

Genetic engineering letter. Published/Created: Washington, D.C.: Gershon W. Fishbein, [1981- Description: 12 v.; 28 cm. Vol. 1, no. 1 (Jan. 10, 1981)- Ceased with v. 12, published 1992. Current Frequency: Twice a month Continued by: Biotech daily 1067-1196 (DLC) 92650248 (OCoLC)26901088 ISSN: 0276-1882 Cancel/Invalid LCCN: sn 81004235 CODEN: GELEEV Notes: Title from caption. SERBIB/SERLOC merged record Indx'd selectively by: Bibliography of agriculture 0006-1530 BioBusiness 1984- Subjects: Genetic engineering--Periodicals. Biotechnology--Periodicals. Biotechnology--periodicals. Genetic Engineering--periodicals. LC Classification: QH442 .G453 Dewey Class No.: 660/.6/05 19

Genetic engineering of microorganisms / edited by Alfred Pühler. Published/Created: Weinheim; New York: VCH, c1993. Related Authors: Pühler, A. Description: viii, 178 p.: ill.; 24 cm. ISBN: 3527300392 (Weinheim: acid-free paper) 1560818263 (New York: acid-free paper) Notes: Includes bibliographical references and index. Subjects: Microbial genetic engineering. Microbial genetics. Microbial biotechnology. LC Classification: TP248.6 .G4617 1993 Dewey Class No.: 660/.62 20

Genetic engineering of plants: a technology impact report. Published/Created: New York, N.Y. (106 Fulton St., New York 10038): Frost & Sullivan, c1989. Related Authors: Frost & Sullivan. Description: 235 p.; 29 cm. Notes: "T008." "Spring 1989." Includes bibliographical references (p. 212-214). Subjects: Agricultural biotechnology. Plant biotechnology. Plant genetic engineering. LC Classification: IN PROCESS (ONLINE)

Genetic engineering with plant viruses / edited by T. Michael A. Wilson, Jeffrey W. Davies. Published/Created: Boca Raton: CRC Press, c1992. Related Authors: Wilson, T. Michael A. Davies, Jeffrey W. Description: 361 p.: ill.; 25 cm. ISBN: 0849368014 (alk. paper) Notes: Includes bibliographical references and index. Subjects: Plant genetic engineering. Plant viruses--Biotechnology. LC Classification: QK981.5 .G47 1992 Dewey Class No.: 576/.6483 20

Genetic engineering, human genetics, and cell biology: evolution of technological issues. Biotechnology (supplemental report III): report / prepared for the Subcommittee on Science, Research, and Technology of the Committee on Science and Technology, U.S. House of Representatives, Ninety-sixth Congress, second session by the Science Policy Research Division, Congressional Research Service, Library of Congress. Published/Created: Washington: U.S. G.P.O.: for sale by the Supt. of Docs., U.S. G.P.O., 1980. Related Authors: United States. Congress. House. Committee on Science and Technology. Subcommittee on Science, Research, and Technology. Library of Congress. Science Policy Research Division. Related Titles: Biotechnology. Description: xvii, 144 p.: ill.; 24 cm. Notes: At head of Committee print. "Serial DDD." "August 1980." S/N 052-070-054417 Item 1025-A Subjects: Genetic engineering--Research--United States. Genetic engineering--Government policy--United States. Human genetics--Research--United States. Human genetics--Government policy--United States. Cytology--Research--United States. Cytology--Government policy--United States. Biotechnology--Research--United States. Biotechnology--Government policy--United States. LC Classification: QH442 .G448 1980 Dewey Class No.: 574.87/322 19

Genetic manipulation: techniques and applications / edited by J.M. Grange, A. Fox & N.L. Morgan. Published/Created: Oxford; Boston: Blackwell Scientific Publications, 1991. Related Authors: Grange, John M. Fox, A. (Arnold), 1927- Morgan, N. L. (Neil L.), 1951- Society for Applied Bacteriology. Autumn Meeting (1989: South Bank Polytechnic) Description: xiii, 401 p.: ill.; 24 cm. ISBN: 0632029269 Notes: "Based on the Society of Applied Bacteriology Autumn Meeting held at the South Bank Polytechnic, London, on 18 October 1989"--Pref. Includes index. Includes bibliographical references. Subjects: Genetic engineering--Congresses. Biotechnology--Congresses. Biotechnology--congresses. Genetic Engineering--congresses. Series: Technical series (Society for Applied Bacteriology); no. 28. Variant Series: The Society for Applied Bacteriology technical series; no. 28 LC Classification: TP248.6 .G465 1990 Dewey Class No.: 660/.65 20

Genetic technology: a guide to key R & D projects. Published/Created: Fort Lee, N.J., USA: Technical Insights, Inc., c1983. Related Authors: Technical Insights, Inc. Description: vii, 147 leaves; 29 cm. ISBN: 0914993011 (pbk.) Notes: Includes index. Subjects: Genetic engineering--Research--Directories. Genetic engineering industries--Directories. Biotechnology--Research--Directories. LC Classification: TP248.6 .G47 1983 Dewey Class No.: 660/.6 19

Genetic technology: a guide to key R & D projects. Edition Information: Rev., updated, expanded. Published/Created: Fort Lee, N.J., USA: Technical Insights, Inc., c1984. Related Authors: Technical Insights, Inc. Description: vii, 280 leaves; 29 cm. ISBN: 0914993100 (pbk.) Notes: Includes index. Subjects: Genetic engineering--Research--Directories. Genetic engineering industries--Directories. Biotechnology--Research--Directories. LC

Classification: TP248.6 .G47 1984
Dewey Class No.: 660/.6 19

Genetically engineered marine organisms: environmental and economic risks and benefits / edited by Raymond A. Zilinskas and Peter J. Balint. Published/Created: Boston: Kluwer Academic Publishers, c1998. Related Authors: Zilinskas, Raymond A. Balint, Peter J., 1950- Description: xvi, 231 p.: ill. (some col.); 25 cm. ISBN: 0412152517 (alk. paper) Notes: Includes bibliographical references and index. Subjects: Marine biotechnology. Transgenic organisms. LC Classification: TP248.27.M37 G46 1998 Dewey Class No.: 333.95/615 21

Genetics and biotechnology of lactic acid bacteria / edited by Michael J. Gasson, and Willem M. De Vos. Edition Information: 1st ed. Published/Created: London; New York: Blackie Academic & Professional, 1994. Related Authors: Gasson, Michael J. Vos, Willem M. de. Description: xi, 300 p.: ill.; 24 cm. ISBN: 075140098X Notes: Includes bibliographical references and index. Subjects: Lactic acid bacteria. Microbial genetic engineering. Microbial biotechnology. LC Classification: QR121 .G46 1994 Dewey Class No.: 589.9/5 20

Genetics and product formation in streptomyces / edited by Simon Baumberg, Hans Krügel, and Dieter Noack. Published/Created: New York: Plenum Press, c1991. Related Authors: Baumberg, S. Krügel, Hans. Noack, Dieter. Federation of European Microbiological Societies. Description: xi, 328 p.: ill.; 26 cm. ISBN: 0306438852 Notes: "Proceedings of a symposium held under the auspices of the Federation of European Microbiological Societies, held May 1-6, 1990, in Erfurt, Germany"--T.p. verso. Includes bibliographical references and index. Subjects: Streptomyces--Congresses. Bacterial genetics--Congresses. Microbial metabolism--Congresses. Microbial biotechnology--Congresses. Series: The Language of science FEMS symposium; no. 55. Variant Series: Federation of European Microbiological Societies symposium series; no. 55 LC Classification: QR82.S8 G46 1991 Dewey Class No.: 589.9/2 20

Genomics and proteomics technologies: 22-23 January 2001, San Jose, USA / Ramesh Raghavachari, Weihong Tan, chairs/editors; sponsored ... by SPIE--the International Society for Optical Engineering. Published/Created: Bellingham, Wash., USA: SPIE, c2001. Related Authors: Raghavachari, Ramesh. Tan, Weihong. Society of Photo-optical Instrumentation Engineers. Description: v, 82 p.: ill.; 28 cm. ISBN: 0819439428 Notes: Errata pages inserted. Includes bibliographical references and index. Subjects: Genetic engineering--Congresses. Genomes--Congresses. Molecular genetics--Congresses. Proteins--Analysis--Congresses. Biotechnology--Congresses. Series: Progress in biomedical optics and imaging, 1605-7422; vol. 2, no. 21 Proceedings of SPIE--the International Society for Optical Engineering; v. 4264. Variant Series: Proceedings of SPIE; v. 4264 LC Classification: TP248.6 .G523 2001 Dewey Class No.: 660.6/5 21

Gray, N. F. Biology of wastewater treatment / N.F. Gray. Published/Created: Oxford [England]; New York: Oxford

University Press, 1989. Description: xiv, 828 p.: ill.; 25 cm. ISBN: 0198590148: Notes: Includes bibliographical references (p. [729]-814) and index. Subjects: Sewage--Purification--Biological treatment. Sewage disposal. Biotechnology. LC Classification: TD755 .G73 1989 Dewey Class No.: 628.3/51 19

Gros, Francois, 1925- The gene civilization / Francois Gros. Published/Created: New York: McGraw Hill, 1992. Description: 136 p.; 23 cm. ISBN: 0070249636: Notes: Translation of: La civilisation du gène. Includes bibliographical references (p. 135-136). Subjects: Molecular genetics. Human molecular genetics. Biotechnology. Series: McGraw-Hill horizons of science series LC Classification: QH442 .G7913 1992 Dewey Class No.: 574.87/328 20

Gross, Cynthia S. The new biotechnology: putting microbes to work / Cynthia S. Gross. Published/Created: Minneapolis: Lerner Publications, c1988. Description: 96 p.: ill. (some col.); 23 cm. ISBN: 0822515830 (lib. bdg.): Summary: Describes the new field of biotechnology and the potential and dangers involved in its development. Notes: Includes index. Subjects: Biotechnology--Juvenile literature. Biotechnology. LC Classification: TP248.2 .G76 1988 Dewey Class No.: 660/.6 19

Haberman, Allan B. Cellular and tissue engineering / Allan B. Haberman. Published/Created: Waltham, Mass.: Decision Resources, Inc., [1996] Related Authors: Decision Resources, Inc. Description: 2 v.: ill.; 30 cm. Contents: pt. 1. Applications for treatment of central nervous system diseases -- pt. 2. Applications for treatment of diabetes. Notes: "September 1996." Subjects: Nervous system--Degeneration--Treatment. Nervous system--Transplantation. Animal cell biotechnology. Diabetes--Treatment. Pancreas--Transplantation. Series: DR reports LC Classification: RC394.D35 H33 1996

Hall, Stephen S. Invisible frontiers: the race to synthesize a human gene / Stephen S. Hall. Published/Created: Redmond, Wash.: Tempus Books of Microsoft Press; [New York]

Hochfeld, William L. Building blocks: biotechnology reagents and consumables / William L. Hochfeld. Published/Created: Buffalo Grove, IL: Interpharm Press, c1994. Description: xvi, 296 p.: ill.; 24 cm. ISBN: 0935184597 Notes: Includes bibliographical references (p. 281-293) and index. Subjects: Biotechnology--Technique. Biological reagents. Series: The Interpharm biotech 2000 series; v. 1 LC Classification: TP248.24 .H63 1994 Dewey Class No.: 660/.6/028 20

Holland, H. L. (Herbert L.) Organic synthesis with oxidative enzymes / H.L. Holland. Published/Created: New York: VCH, c1992. Description: x, 463 p.: ill.; 24 cm. ISBN: 0895737795 (VCH Publishers, Inc.) 3527279563 (VCH Verlagsgesellschaft) Notes: Includes bibliographical references and index. Subjects: Enzymes--Biotechnology. Oxidases--Industrial applications. LC Classification: TP248.65.E59 H65 1991 Dewey Class No.: 660/.634 20

Human gene therapy: technology analysis and market forecast / Lynn Gray, project analyst. Published/Created: Norwalk, CT: Business Communications Co., c1998. Related Authors: Kelly, Paul E. Business Communications Co. Description: xxi, 293 p.: ill.; 28 cm. ISBN: 1569653895 Subjects: Genetic engineering industry--United States. Biotechnology industries--United States. Market surveys--United States. Series: Business opportunity report; C-160R LC Classification: HD9999.G453 U645 1998

Human gene therapy: technology analysis and market forecast / Paul E. Kelly, project analyst. Published/Created: Norwalk, CT: Business Communications Co., c1993. Related Authors: Kelly, Paul E. Business Communications Co. Description: xvi, 219 leaves: ill.; 28 cm. ISBN: 0893369551 Subjects: Genetic engineering industry--United States. Biotechnology industries--United States. Market surveys--United States. Series: Business opportunity report; C-160 LC Classification: HD9999.G453 U645 1993

Hydrocarbons in biotechnology: proceedings of a meeting / organized by the Institute of Petroleum and held at the University of Kent at Canterbury, UK, 25-26 September 1979; edited by D.E.F. Harrison, I.J. Higgins and R. Watkinson. Published/Created: London; Philadelphia, Pa.: Heyden, c1980. Related Authors: Harrison, D. E. F. (David Ernest Forester), 1939- Higgins, I. J. (Irving John) Watkinson, R. J. Institute of Petroleum (Great Britain) Description: viii, 201 p.: ill.; 24 cm. ISBN: 0855013257 Notes: Includes bibliographies and index. Subjects: Biotechnology--Congresses. Hydrocarbons--Industrial applications--Congresses. LC Classification: TP248.3 .H9 1980 Dewey Class No.: 661/.81 19

Immobilization of enzymes and cells / edited by Gordon F. Bickerstaff. Published/Created: Totowa, N.J.: Humana Press, c1997. Related Authors: Bickerstaff, Gordon F. Description: xiv, 367 p.: ill.; 24 cm. ISBN: 0896033864 (alk. paper) Notes: Includes bibliographical references and index. Subjects: Immobilized enzymes--Biotechnology. Immobilized cells--Biotechnology. Enzymes, Immobilized. Cells, Immobilized. Biotechnology--methods. Series: Methods in biotechnology; 1 LC Classification: TP248.65.I45 I46 1997 Dewey Class No.: 660/.634 20

In vitro cultivation of animal cells. Published/Created: Oxford: Butterworth-Heinemann on behalf of Open universiteit and University of Greenwich (formerly Thames Polytechnic), 1993. Related Authors: Open Universiteit (Heerlen, Netherlands) University of Greenwich. Description: ix, 264 p., [16] p.: ill.; 25 cm. ISBN: 0750605553 (pbk) Notes: Includes index. "Suggestions for further reading": p. 263-264. Subjects: Animal cell biotechnology. Cell culture. Animals cells biotechnology Series: Biotechnology by Open Learning (Series) Variant Series: BIOTOL, Biotechnology by Open Learning LC Classification: TP248.27.A53 I52 1993 Dewey Class No.: 660/.6 20

In vitro cultivation of plant cells. Published/Created: Oxford [England]: Butterworth-Heinemann, 1993. Related Authors: BIOTOL (Project) Open Universiteit (Heerlen, Netherlands) University of Greenwich. Description:

x, 200 p.: ill.; 25 cm. ISBN: 0750605545 Notes: "Published on behalf of Open Universiteit and University of Greenwich." GB92-03014 Includes bibliographical references (p. 195-196) and index. Subjects: Plant cell culture--Programmed instruction. Plant tissue culture--Programmed instruction. Plant biotechnology--Programmed instruction. Series: Biotechnology by Open Learning (Series) Variant Series: BIOTOL, Biotechnology by Open Learning LC Classification: QK725 .I445 1993 Dewey Class No.: 581/.0724 20

Industrial biotechnological polymers / edited by Charles G. Gebelein, Charles E. Carraher, Jr. Published/Created: Lancaster, Pa.: Technomic Pub. Co., c1995. Related Authors: Gebelein, Charles G. Carraher, Charles E. Description: xiv, 408 p.: ill.; 23 cm. ISBN: 1566762928 Notes: Includes bibliographical references. Subjects: Polymers--Biotechnology. LC Classification: TP248.65.P62 I53 1995

International industrial biotechnology. Published/Created: Swansea: Biotechnology Centre Wales, 1986-c1989. Related Authors: Biotechnology Centre Wales. Description: 4 v.: ill., ports.; 28 cm. Vol. 6, issue 2 (Feb./Mar. 1986)-v. 9, issue 6 (Nov./Dec. 1989). Current Frequency: Bimonthly Continues: Industrial biotechnology 0268-3024 Continued by: Genetic engineer & biotechnologist 0959-020X (DLC)sn 90018282 (OCoLC)21832452 ISSN: 0269-7815 CODEN: IIBIE9 Notes: Title from cover. Imprint varies: Cambridge; New York, NY: Cambridge University Press, -Nov./Dec. 1989. SERBIB/SERLOC merged record Indx'd selectively by: BioBusiness

1989- Chemical abstracts 0009-2258 - 1989 Subjects: Biotechnology--Periodicals. Biotechnology industries--Periodicals. Biotechnology--periodicals. LC Classification: TP248.13 .I63 Dewey Class No.: 660/.6 19

International Symposium on Biotechnology (1990: Bratislava, Czechoslovakia) Environmental biotechnology: proceedings of the International Symposium on Biotechnology, Bratislava, Czecho-Slovakia, June 27-29, 1990 / edited by A. Blazej and V. Prívarová. Published/Created: Amsterdam; New York: Elsevier, 1991. Related Authors: Blazej, Anton. Prívarová, V. Description: 444 p.: ill.; 25 cm. ISBN: 0444987207 Notes: Includes bibliographical references. Subjects: Bioremediation--Congresses. Microbial biotechnology--Congresses. Biotechnology--Environmental aspects--Congresses. Biomass energy--Environmental aspects--Congresses. Series: Studies in environmental science; 42 LC Classification: TD192.5 .I57 1990 Dewey Class No.: 628 20

Intrabodies: basic research and clinical gene therapy applications / Wayne A. Marasco, ed. Published/Created: Berlin; New York: Landes Bioscience, c1998. Related Authors: Marasco, Wayne A., 1953- Description: 211 p.: ill.; 24 cm. ISBN: 3540641513 (acid-free paper) Notes: Includes bibliographical references and index. Subjects: Immunoglobulins--Biotechnology. Gene targeting. Gene therapy. Cellular signal transduction. Antibodies--physiology. Signal Transduction--physiology. Gene Expression--physiology. Series: Molecular biology intelligence unit LC Classification: TP248.65.I49 I56 1998

Dewey Class No.: 616.07/98 21

Intracellular ribozyme applications: principles and protocols / edited by John J. Rossi and Larry A. Couture. Published/Created: Wymondham, Norfolk, England: Horizon Scientific Press, c1999. Related Authors: Rossi, John J. Couture, Larry A. Description: x, 294 p.: ill.; 25 cm. ISBN: 1898486174 Notes: Includes bibliographical references and index. Subjects: Catalytic RNA. Catalytic RNA--Therapeutic use. Genetic regulation. RNA, Catalytic--genetics. Biotechnology. RNA, Catalytic--metabolism. Cytology. Catalytic RNA. LC Classification: QP623.5.C36 I58 1999 Dewey Class No.: 572.8/8 21

Introduction to biocatalysis using enzymes and micro-organisms / Stanley M. Roberts ... [et al]. Published/Created: Cambridge; New York: Cambridge University Press, 1995. Related Authors: Roberts, Stanley M. Description: xii, 195 p.: ill.; 24 cm. ISBN: 0521430704 (hardback) 0521436850 (pbk.) Notes: Includes bibliographical references and index. Subjects: Microbial biotechnology. Enzymes--Biotechnology. Biotransformation (Metabolism) LC Classification: TP248.27.M53 I57 1995 Dewey Class No.: 660/.63 20

Jones, H. Michelle. Cell culture systems and conventional bioreactor technology / H. Michelle Jones, project analyst. Published/Created: Norwalk, CT: Business Communications Co., c1997. Related Authors: Business Communications Co. Description: 1 v. (various pagings): ill.; 28 cm. ISBN: 1569653828 Subjects: Biotechnology industries--United States. Bioreactors--

United States. Market surveys--United States. Series: Business opportunity report; C-086R LC Classification: HD9999.B443 U645 1997

Juma, Calestous. Genetic resources and biotechnology in Kenya: towards long-term food security / Calestous Juma. Published/Created: Nairobi: Public Law Institute, [1987] Related Authors: Public Law Institute (Kenya) Description: 124 leaves; 30 cm. Notes: "A public policy study." "July 1987." Bibliography: leaves 110-115. Subjects: Crops--Kenya--Germplasm resources. Plant biotechnology--Kenya. Germplasm resources, Plant--Law and legislation--Kenya. Biotechnology. LC Classification: SB123.3 .J86 1987

Krul, Kenneth G. Strategic approaches to developing and commercializing gene therapy products / Kenneth G. Krul. Published/Created: Waltham, MA: Decision Resources, c1994. Related Authors: Decision Resources Corporation. Description: xi, 90 p.: ill.; 28 cm. Notes: "R940402"--Cover. Includes bibliographical references. Subjects: Genetic engineering industry--United States. Biotechnology industries--United States. Market surveys--United States. Series: DR reports LC Classification: HD9999.G453 U657 1994

Krul, Kenneth G. Strategic approaches to developing and commercializing gene therapy products / Kenneth G. Krul. Published/Created: Waltham, MA: Decision Resources, c1994. Related Authors: Decision Resources Corporation. Description: xi, 90 p.: ill.; 28 cm. Notes: "R940402"--Cover. Includes bibliographical references. Subjects: Genetic engineering industry--

United States. Biotechnology industries--United States. Market surveys--United States. Series: DR reports LC Classification: HD9999.G453 U657 1994

Krul, Kenneth G. The commercial prospects for gene therapy technology / Kenneth D. [i.e. G.] Krul. Published/Created: Waltham, Mass.: Decision Resources, [1998] Related Authors: Decision Resources Corporation. Description: 156 p.; 30 cm. Notes: "February 1998." Subjects: Genetic engineering industry. Genetic engineering--Research--Directories. Biotechnology industries. Biotechnology--Research--Directories. Market surveys. Series: DR reports LC Classification: HD9999.G452 K78 1998

Krul, Kenneth G. The commercial prospects for gene therapy technology / Kenneth D. [i.e. G.] Krul. Published/Created: Waltham, Mass.: Decision Resources, [1998] Related Authors: Decision Resources Corporation. Description: 156 p.; 30 cm. Notes: "February 1998." Subjects: Genetic engineering industry. Genetic engineering--Research--Directories. Biotechnology industries. Biotechnology--Research--Directories. Market surveys. Series: DR reports LC Classification: HD9999.G452 K78 1998

Krupp, Guido, 1952- Ribozyme: biochemistry and biotechnology / Guido Krupp, Rajesh K. Gaur. Published/Created: Natick, MA: Eaton Pub., c2000. Related Authors: Gaur, Rajesh K., 1961- Description: x, 514 p.: ill.; 26 cm. ISBN: 1881299287 Notes: Includes bibliographical references and index. Subjects: Catalytic RNA. Catalytic RNA--Biotechnology. LC Classification: QP623.5.C36 K78 2000

Dewey Class No.: 572/.788 21

Krupp, Guido, 1952- Ribozyme: biochemistry and biotechnology / Guido Krupp, Rajesh K. Gaur. Published/Created: Natick, MA: Eaton Pub., c2000. Related Authors: Gaur, Rajesh K., 1961- Description: x, 514 p.: ill.; 26 cm. ISBN: 1881299287 Notes: Includes bibliographical references and index. Subjects: Catalytic RNA. Catalytic RNA--Biotechnology. LC Classification: QP623.5.C36 K78 2000 Dewey Class No.: 572/.788 21

Laing, E., 1931- The new biology: new hope, new threat, or new dilemmas / E. Laing. Published/Created: Accra: Ghana Universities Press, 1989. Description: 39 p.; 21 cm. ISBN: 9964301707 Notes: Includes bibliographical references. Subjects: Bioethics. Genetic Engineering. Genetics. Bioethics. Biotechnology. Genetic engineering. Science--Moral and ethical aspects. LC Classification: QH332 .L35 1989

Laing, E., 1931- The new biology: new hope, new threat, or new dilemmas / E. Laing. Published/Created: Accra: Ghana Universities Press, 1989. Description: 39 p.; 21 cm. ISBN: 9964301707 Notes: Includes bibliographical references. Subjects: Bioethics. Genetic Engineering. Genetics. Bioethics. Biotechnology. Genetic engineering. Science--Moral and ethical aspects. LC Classification: QH332 .L35 1989

Lalithakumari, D. Fungal protoplast: a biotechnological tool / D. Lalithakumari. Published/Created: Enfield, NH: Science Pub., c2000. Description: xii, 184 p.: ill. (some col.);

24 cm. ISBN: 1578080932 Notes: Includes bibliographical references (p. [158]-184). Subjects: Fungi-- Biotechnology. Fungal protoplasts. LC Classification: TP248.27.F86 L35 2000 Dewey Class No.: 660.6 21

Lancini, Giancarlo. Biotechnology of antibiotics and other bioactive microbial metabolites / Giancarlo Lancini and Rolando Lorenzetti. Published/Created: New York: Plenum Press, c1993. Related Authors: Lorenzetti, Rolando. Description: viii, 236 p.: ill.; 24 cm. ISBN: 0306446030 Notes: Includes bibliographical references and index. Subjects: Antibiotics--Biotechnology. Microbial metabolites--Biotechnology. Microbial biotechnology. LC Classification: TP248.65.A57 L36 1993 Dewey Class No.: 615/.329 20

Lancini, Giancarlo. Biotechnology of antibiotics and other bioactive microbial metabolites / Giancarlo Lancini and Rolando Lorenzetti. Published/Created: New York: Plenum Press, c1993. Related Authors: Lorenzetti, Rolando. Description: viii, 236 p.: ill.; 24 cm. ISBN: 0306446030 Notes: Includes bibliographical references and index. Subjects: Antibiotics--Biotechnology. Microbial metabolites--Biotechnology. Microbial biotechnology. LC Classification: TP248.65.A57 L36 1993 Dewey Class No.: 615/.329 20

Lanzavecchia, Giuseppe. The impact of biotechnology on working conditions / [by Guiseppe [sic] Lanzavecchia, Danielle Mazzonis]. Published/Created: Loughlinstown House, Shankill, Co. Dublin, Ireland: European Foundation for the Improvement of Living and Working Conditions, c1987. Related Authors: Gattegno Mazzonis, Danielle.

Description: x, 105 p.; 30 cm. ISBN: 9282567672 Notes: Includes bibliographical references (p. 71-83). Subjects: Work environment. Industrial hygiene. Biotechnology industries. Biotechnology. Series: Research report (European Foundation for the Improvement of Living and Working Conditions) Variant Series: Research report LC Classification: HD6955 .L32 1987 Dewey Class No.: 331.25 20

Large, Peter J. Methylotrophy and biotechnology / Peter J. Large and Charles W. Bamforth. Published/Created: Harlow, Essex, England: Longman Scientific & Technical; New York, N.Y.: Wiley, 1988. Related Authors: Bamforth, Charles W., 1952- Description: 303 p.: ill.; 25 cm. ISBN: 047021144X (Wiley): Notes: Includes index. Bibliography: p. 245-300. Subjects: Methylotrophic microorganisms--Biotechnology. LC Classification: TP248.27.M47 L37 1988

Large, Peter J. Methylotrophy and biotechnology / Peter J. Large and Charles W. Bamforth. Published/Created: Harlow, Essex, England: Longman Scientific & Technical; New York, N.Y.: Wiley, 1988. Related Authors: Bamforth, Charles W., 1952- Description: 303 p.: ill.; 25 cm. ISBN: 047021144X (Wiley): Notes: Includes index. Bibliography: p. 245-300. Subjects: Methylotrophic microorganisms--Biotechnology. LC Classification: TP248.27.M47 L37 1988 Dewey Class No.: 660/.62 19

Lipid biotechnology / edited by Tsung Min Kuo, Harold Gardner. Published/Created: New York: Marcel Dekker, c2002. Projected Pub. Date: 0202 Related Authors: Kuo, Tsung Min,

1942- Gardner, Harold W., 1935-
Description: p. cm. ISBN: 0824706196
(alk. paper) Notes: Includes index.
Subjects: Lipids--Biotechnology.
Lipids--Synthesis. LC Classification:
TP248.65.L57 L565 2002 Dewey Class
No.: 660.6/3 21

Lipid synthesis and manufacture / edited by
Frank D. Gunstone. Published/Created:
Boca Raton: Sheffield Academic Press,
1998. Projected Pub. Date: 9810
Related Authors: Gunstone, F. D.
Description: p. cm. ISBN: 0849397375
(alk. paper) Notes: Includes
bibliographical references and index.
Subjects: Lipids--Biotechnology.
Lipids--Synthesis. Series: The chemistry
and technology of oils and fats LC
Classification: TP248.65.L57 L56 1998
Dewey Class No.: 660.6/3 21

Liposomes as tools in basic research and
industry / edited by Jean R. Philippot
and Francis Schuber.
Published/Created: Boca Raton: CRC
Press, c1995. Related Authors:
Philippot, Jean R. Schuber, Francis.
Description: 277 p.: ill.; 27 cm. ISBN:
0849345693 (acid-free paper) Notes:
Includes bibliographical references and
index. Subjects: Liposomes. Liposomes-
-Biotechnology. LC Classification:
QH601 .L568 1995 Dewey Class No.:
574.87/4 20

Maintaining cultures for biotechnology and
industry / edited by Jennie C. Hunter-
Cevera, Angela Belt.
Published/Created: San Diego:
Academic Press, c1996. Related
Authors: Hunter-Cevera, Jennie C. Belt,
Angela. Description: xiv, 263 p.: ill.; 23
cm. ISBN: 0123619459 (hc: alk. paper)
0123619467 (pbk.: alk. paper) Notes:
Includes bibliographical references and

index. Subjects: Culture media
(Biology) Biotechnology. LC
Classification: TP248.25.C44 M35 1996
Dewey Class No.: 660/.6 20

Mammalian cell biotechnology in protein
production /editors, Hansjörg Hauser,
Roland Wagner. Published/Created:
Berlin; New York: Walter de Gruyter,
1997. Related Authors: Hauser,
Hansjörg, 1949- Wagner, Roland, 1956-
Description: xix, 491 p.: ill.; 25 cm.
ISBN: 3110134039 (alk. paper: cover)
Cancelled ISBN: 03110134039 Notes:
Includes bibliographical references and
index. Subjects: Proteins--
Biotechnology. Animal cell
biotechnology. Mammals--Cytology.
LC Classification: TP248.65.P76 M36
1997 Dewey Class No.: 660/.63 20

Mammalian cell biotechnology: a practical
approach / edited by M. Butler.
Published/Created: Oxford [England];
New York: IRL Press at Oxford
University Press, c1991. Related
Authors: Butler, M. (Michael), 1947-
Description: xix, 245 p.: ill.; 23 cm.
ISBN: 019963209X (pbk.):
0199632073: Notes: Includes
bibliographical references and index.
Subjects: Animal cell biotechnology.
Mammals. Series: The Practical
approach series LC Classification:
TP248.27.A53 M36 1991 Dewey Class
No.: 660/.6 20

Mammalian cell culture technology / editors
Mikio Shikita, Isao Yamane.
Published/Created: Tokyo, Japan: Soft
Science Publications, c1985. Related
Authors: Shikita, Mikio, 1932- Yamane,
Isao, 1922- International Symposium on
Mammalian Cell Culture Technology
(1982: Kyoto, Japan) Description: xvi,
207 p.: ill.; 27 cm. ISBN: 488171001X:

Notes: Based on the proceeding of the International Symposium on Mammalian Cell Culture Technology, held in Kyoto, Japan, 1982. Includes bibliographies and index. Subjects: Cell culture--Congresses. Mammals--Cytology--Congresses. Biotechnology--Congresses. LC Classification: QH585 .M35 1985 Dewey Class No.: 599/.007/24 19

Manipulating secondary metabolism in culture / edited by Richard J. Robins & Michael J.C. Rhodes. Published/Created: Cambridge [Cambridgeshire]; New York: Cambridge University Press, 1988. Related Authors: Robins, Richard J. Rhodes, M. J. C. International Association for Plant Tissue Culture. UK Section. Meeting (2nd: 1987: AFRC Institute of Federal Research, Norwich) Description: x, 312 p.: ill.; 26 cm. ISBN: 0521362547 Notes: "Presentations made to the Second Meeting of the UK Section of the International Association for Plant Tissue Culture ... held at the AFRC Institute of Federal Research, Norwich, on the 16th and 17th September 1987"--Pref. Includes bibliographies and indexes. Subjects: Plant biotechnology--Congresses. Metabolism, Secondary--Regulation--Congresses. Plant cell culture--Congresses. LC Classification: TP248.27.P55 M36 1988 Dewey Class No.: 581/.07/24 19

Manual of industrial microbiology and biotechnology / editors, Arnold L. Demain and Nadine A. Solomon. Published/Created: Washington, D.C.: American Society for Microbiology, 1986. Related Authors: Demain, A. L. (Arnold L.) Solomon, N. A. (Nadine A.) Related Titles: Industrial microbiology and biotechnology. Description: xi, 466 p.: ill.; 29 cm. ISBN: 0914826727: 0914826735 (soft): Notes: Includes bibliographies and indexes. Subjects: Industrial microbiology--Handbooks, manuals, etc. Biotechnology--Handbooks, manuals, etc. LC Classification: QR53 .M33 1986 Dewey Class No.: 660/.6 19

Manual of industrial microbiology and biotechnology / editors-in-chief, Arnold L. Demain, Julian E. Davies; editors, Ronald M. Atlas ... [et al.]. Edition Information: 2nd ed. Published/Created: Washington, D.C.: ASM Press, c1999. Related Authors: Demain, A. L. (Arnold L.), 1927- Davies, Julian E. Atlas, Ronald M., 1946- Description: xv, 830 p.: ill.; 29 cm. ISBN: 1555811280 Notes: Includes bibliographical references and indexes. Subjects: Industrial microbiology--Handbooks, manuals, etc. Industrial microbiology--Handbooks, manuals, etc. Biotechnology--Handbooks, manuals, etc. LC Classification: QR53 .M33 1999 Dewey Class No.: 660.6/2 21

Marine bioprocess engineering: proceedings of an International Symposium organized under auspices of the Working Party on Applied Biocatalysis of the European Federation of Biotechnology and the European Society for Marine Biotechnology, Noordwijkerhout, The Netherlands, November 8-11, 1998 / edited by R. Osinga ... [et al.]. Published/Created: Amsterdam; New York: Elsevier, 1999. Related Authors: Osinga, R. European Federation of Biotechnology. Working Party on Applied Biocatalysis. European Society for Marine Biotechnology. Description: 414 p.: ill.

(some col.); 27 cm. ISBN: 0444503870
Notes: Includes bibliographical
references and indexes. Subjects:
Marine biotechnology--Congresses.
Series: Progress in industrial
microbiology; v. 35 LC Classification:
QR53 .P7 vol. 35 TP248.27.M37
Dewey Class No.: 660.6 21

Marine biotechnology. Published/Created:
New York, NY: Springer-Verlag New
York Inc., c1999- Related Authors:
European Society for Marine
Biotechnology. Marin Baiotekunoroj¯i
Kenky¯ukai (Japan) Description: v.: ill.;
28 cm. Vol. 1, no. 1 (Jan./Feb. 1999)-
Current Frequency: Bimonthly ISSN:
1436-2228 CODEN: MABIFW Notes:
Title from cover. Some vols.
accompanied by supplements. Official
journal of: European Society for Marine
Biotechnology, and: Japanese Society of
Marine Biotechnology. Formed by the
merger of: Journal of marine
biotechnology, and: Molecular marine
biology and biotechnology.

McCullough, Kenneth C. (Kenneth Charles),
1951- Monoclonal antibodies in biology
and biotechnology: theoretical and
practical aspects / Kenneth C.
McCullough and Raymond E. Spier.
Published/Created: Cambridge
[England]; New York: Cambridge
University Press, 1990. Related
Authors: Spier, R. (Raymond)
Description: xi, 387 p.: ill.; 24 cm.
Cancelled ISBN: 0521285901 Notes:
Includes index. Bibliography: p. 352-
379. Subjects: Monoclonal antibodies--
Biotechnology. Series: Cambridge
studies in biotechnology; 8 LC
Classification: TP248.65.M65 M33
1990 Dewey Class No.: 660/.63 19

McKelvey, Maureen D. Evolutionary
innovations: the business of
biotechnology / Maureen D. McKelvey.
Published/Created: New York: Oxford
University Press, 2000. Projected Pub.
Date: 0003 Description: p. cm. ISBN:
0198297246 Notes: Includes
bibliographical references and index.
Subjects: Genentech, Inc. KabiVitrum
Sverige AB. Genetic engineering.
Biotechnology industries--United
States. Biotechnology industries--
Sweden. Recombinant human insulin.
Recombinant human somatotropin. LC
Classification: TP248.6 .M38 2000
Dewey Class No.: 338.7/6606 21

McKelvey, Maureen D. Evolutionary
innovations: the business of
biotechnology / Maureen D. McKelvey.
Published/Created: Oxford; New York:
Oxford University Press, 1996.
Description: xiv, 319 p.: ill.; 24 cm.
ISBN: 0198289960 (acid-free paper)
Notes: Includes bibliographical
references (p. [300]-315) and index.
Subjects: Genetic engineering.
Biotechnology industries--United
States. Biotechnology industries--
Sweden. Recombinant human insulin.
Recombinant human somatotropin. LC
Classification: TP248.6 .M38 1996
Dewey Class No.: 660/.65 20

Mellon, Margaret G. Biotechnology and the
environment: a primer on the
environmental implications of genetic
engineering / Margaret Mellon.
Published/Created: Washington, DC:
National Biotechnology Policy Center,
Environmental Quality Division,
National Wildlife Federation, c1988.
Description: 64 p.: ill.; 28 cm. ISBN:
0912186992: Notes: Bibliography: p.
58-59. Subjects: Genetic engineering--
Industrial applications--Social aspects.

Biotechnology--Environmental aspects. LC Classification: TP248.6 .M45 1988 Dewey Class No.: 363.1 19

Methods in plant molecular biology and biotechnology / edited by Bernard R. Glick, John E. Thompson. Published/Created: Boca Raton: CRC Press, c1993. Related Authors: Glick, Bernard R. Thompson, John E. Description: 360 p., [2] p. of plates: ill. (some col.); 26 cm. ISBN: 0849351642 (acid-free paper) Notes: Includes bibliographical references and index. Subjects: Plant molecular biology--Methodology. Plant biotechnology--Methodology. LC Classification: QK728 .M483 1993 Dewey Class No.: 582/.087328 20

Methods of tissue engineering / edited by Anthony Atala, Robert P. Lanza. Published/Created: San Diego, CA: Academic Press, 2001. Related Authors: Atala, Anthony, 1958- Lanza, R. P. (Robert Paul), 1956- Description: xli, 1285 p.: ill. (some col.); 29 cm. ISBN: 0124366368 (alk. paper) Notes: Includes bibliographical references and index. Subjects: Animal cell biotechnology. Tissue culture. Biomedical engineering. Polymers in medicine. LC Classification: TP248.27.A53 M48 2001 Dewey Class No.: 660.6 21

Mexico '96 Workshop (1996: Cocoyoc, Mexico) Biotechnology for water use and conservation: the Mexico '96 Workshop. Published/Created: Paris: Organisation for Economic Co-operation and Development, c1997. Related Authors: Organisation for Economic Co-operation and Development. Description: 728 p.: ill.; 27 cm. ISBN: 9264155945 (pbk.)

Notes: "The OECD Workshop Mexico '96 on Biotechnology for Water Use and Conservation took place in Cocoyoc, Morelos, on 20-23 October 1996"--Foreword. Includes bibliographical references. Introductory matter also in French. Subjects: Water--Purification--Biological treatment--Congresses. Water quality--Congresses. Water quality management--Congresses. Biotechnology--Congresses. Series: OECD documents LC Classification: TD475 .M48 1996 Dewey Class No.: 628.1/68 21

Micro-algal biotechnology / edited by Michael A. Borowitzka, Lesley J. Borowitzka. Published/Created: Cambridge [Cambridgeshire]; New York: Cambridge University Press, 1988. Related Authors: Borowitzka, Michael A. Borowitzka, Lesley J. Description: x, 477 p.: ill.; 26 cm. ISBN: 0521323495 Notes: Includes bibliographies and indexes. Subjects: Algae culture. Algae--Biotechnology. Algae products. LC Classification: SH389 .M53 1988 Dewey Class No.: 635.9/393 19

"Microbial asepcts [sic] of biotechnology": proceedings of a joint meeting held at the Royal Irish Academy, 27th April 1981 / sponsored by Royal Irish Academy, National Board for Science and Technology, Society for General Microbiology (Irish Branch), and arranged under the auspices of Royal Irish Academy [and] National Commission for Microbiology. Published/Created: [Dublin: Royal Irish Academy, 1982] Related Authors: Royal Irish Academy. Ireland. National Board for Science and Technology. Society for General Microbiology. Irish Branch. Ireland. National Commission

for Microbiology. Related Titles:
"Microbial aspects of biotechnology."
Description: 119 p.: ill.; 30 cm. Notes:
Includes bibliographies. Subjects:
Microbial biotechnology--Congresses.
Industrial microbiology--Congresses.
Fermentation--Congresses. Genetic
engineering--Congresses. LC
Classification: TP248.14 .M53 1982
Dewey Class No.: 660/.62 19

Microbial biomass proteins / edited by
Murray Moo-Young and Kenneth F.
Gregory. Published/Created: London;
New York: Elsevier Applied Science;
New York, NY, USA: Sole distributor
in the USA and Canada, Elsevier
Science Pub. Co., c1986. Related
Authors: Moo-Young, Murray. Gregory,
Kenneth F. Description: xiv, 185 p.: ill.;
23 cm. ISBN: 1851660852: Notes:
Includes bibliographies. Subjects:
Single cell proteins--Biotechnology. LC
Classification: TP248.65.S56 M52 1986
Dewey Class No.: 660/.62 20

Microbial degradation of natural products /
edited by Günther Winkelmann.
Published/Created: Weinheim; New
York: VCH, c1992. Related Authors:
Winkelmann, Günther. Description: xii,
420 p.: ill.; 25 cm. ISBN: 3527283544
(Weinheim: acid-free paper)
1560811692 (New York: acid-free
paper) Notes: Includes bibliographical
references and index. Subjects:
Microbial biotechnology. Microbial
metabolism. Microbial enzymes. LC
Classification: TP248.27.M53 M53
1992 Dewey Class No.: 660/.62 20

Microbial enzymes and biotechnology /
edited by William M. Fogarty.
Published/Created: London; New York:
Applied Science Publishers; New York,
NY, USA: Sole distributor in the USA

and Canada, Elsevier Science Pub. Co.,
c1983. Related Authors: Fogarty,
William M. Description: xiii, 382 p.:
ill.; 23 cm. ISBN: 0853341850
Contents: Microbial amylases / W.M.
Fogarty -- Glucose-transforming
enzymes / C. Bucke -- Pectic enzymes /
W.M. Forgarty and C.T. Kelly --
Microbial cellulases / Tor-Magnus Enari
-- Extracellular microbial lipases / A.R.
Macrae -- Proteinases / Owen P. Ward -
- Enzyme synthesis / F.G. Priest. Notes:
Includes bibliographies and index.
Subjects: Microbial enzymes--
Biotechnology. Microbial enzymes--
Industrial applications. LC
Classification: QR90 .M534 1983
Dewey Class No.: 660/.63 19

Microbial enzymes and biotechnology /
edited by William M. Fogarty &
Catherine T. Kelly. Edition Information:
2nd ed. Published/Created: London;
New York: Elsevier Applied Science,
c1990. Related Authors: Fogarty,
William M. Kelly, Catherine T.
Description: x, 472 p.: ill.; 25 cm.
ISBN: 1851664866 Notes: Includes
bibliographical references and index.
Subjects: Microbial enzymes--
Biotechnology. Microbial enzymes--
Industrial applications. Series: Elsevier
applied biotechnology series LC
Classification: TP248.65.E59 M53 1990
Dewey Class No.: 660/.634 20

Microbial genetic engineering and enzyme
technology / edited by C.P. Hollenberg
and H. Sahm. Published/Created:
Stuttgart; New York: G. Fischer, 1987.
Related Authors: Hollenberg, C. P.
(Cornelis P.) Sahm, H. (Hermann)
Description: vi, 142 p.: ill.; 24 cm.
ISBN: 3437305506: 0895742462 (NY)
Notes: Includes bibliographical
references. Subjects: Genetic

engineering. Microbial genetics.
Industrial microbiology. Enzymes--
Industrial applications. Biotechnology.
Series: Biotec, 0931-1408; 1 LC
Classification: TP248.6 .M53 1987
Dewey Class No.: 660/.62 19

Microbial processes for bioremediation /
edited by Robert E. Hinchee and Fred J.
Brockman, Catherine M. Vogel.
Published/Created: Columbus: Battelle
Press, c1995. Related Authors: Hinchee,
Robert E. Brockman, Fred J., 1957-
Vogel, Catherine M., 1959-
International Symposium on In Situ and
On-Site Bioreclamation (3rd: 1995: San
Diego, Calif.) Description: x, 361 p.:
ill.; 24 cm. ISBN: 1574770098 (hc: alk.
paper) Notes: Papers derived from the
Third International In Situ and On-Site
Bioreclamation Symposium, held in San
Diego, Calif., Apr. 1995. Includes
bibliographical references and index.
Subjects: Bioremediation--Congresses.
Microbial biotechnology--Congresses.
Series: Bioremediation; 3(8) LC
Classification: TD192.5 .M325 1995
Dewey Class No.: 628.5/2 20

Microbial reagents in organic synthesis /
edited by Stefano Servi.
Published/Created: Dordrecht; Boston:
Kluwer Academic Publishers, c1992.
Related Authors: Servi, Stefano, 1943-
Description: x, 490 p.: ill.; 25 cm.
ISBN: 0792319532 (acid-free) Notes:
"Proceedings of the NATO Advanced
Research Workshop Microbial Reagents
in Organic Synthesis, Sestri Levante,
Italy, March 23-27, 1992"--T.p. vers.
"Published in cooperation with NATO
Scientific Affairs Division." Includes
bibliographical references and index.
Subjects: Organic compounds--
Biotechnology--Congresses. Organic
compounds--Synthesis--Congresses.

Microbial metabolism--Congresses.
Microbial biotechnology--Congresses.
Series: NATO ASI series. Series C,
Mathematical and physical sciences; no.
381. Variant Series: NATO ASI series.
Series C, Mathematical and physical
sciences; vol. 381 LC Classification:
TP248.65.O73 M5 1992 Dewey Class
No.: 660/.63 20

Microbial technology in the developing
world / edited by E.J. Da Silva ... [et
al.]. Published/Created: Oxford; New
York: Oxford University press, 1987.
Related Authors: DaSilva, E. J.
Description: viii, 444 p.: ill.; 25 cm.
ISBN: 0198547196: Notes: Includes
bibliographies and index. Subjects:
Biotechnology--Developing countries.
LC Classification: TP248.195.D48 M53
1987 Dewey Class No.: 660/.62/091724
19

Microbiological quality assurance: a guide
towards relevance and reproducibility of
inocula / edited by Michael R.W.
Brown, Peter Gilbert.
Published/Created: Boca Raton, Fl.:
CRC Press, 1995. Related Authors:
Brown, Michael R. W. (Michael Robert
Withington) 1931- Gilbert, Peter.
Description: 299 p.: ill.; 24c m. ISBN:
0849347521 Notes: Includes
bibliographical references and index.
Subjects: Microbial biotechnology.
Microbial inoculants--Quality control.
LC Classification: TP248.27.M53 M54
1995 Dewey Class No.: 660/.62 20

Microbiological research.
Published/Created: Jena: G. Fischer,
c1994- Description: v.: ill.; 28 cm. Vol.
149, 1 (Apr. 1994)- Current Frequency:
Four no. a year Continues: Zentralblatt
für Mikrobiologie 0232-4393 (DLC)
82645751 (OCoLC)8440321 ISSN:

0944-5013 Cancel/Invalid LCCN: sn 94026907 CODEN: MCRSEJ Notes: Title from cover. English and German. SERBIB/SERLOC merged record Indx'd selectively by: Chemical abstracts 0009-2258 Index medicus 0019-3879 1994- Subjects: Microbiology--Periodicals. Soil microbiology--Periodicals. Microbial ecology--Periodicals. Biotechnology--Periodicals. Biotechnology--periodicals. Environmental Microbiology--periodicals. Microbiology--periodicals. LC Classification: QR51 .Z45 Dewey Class No.: 576/.16 19

Microsystem technology: a powerful tool for biomolecular studies / edited by J.M. Köhler, T. Mejevaia, H.P. Saluz. Published/Created: Basel, Switzerland: Boston: Birkhäuser Verlag, c1999. Related Authors: Köhler, J. M. (J. Michael), 1956- Mejevaia, T. Saluz, H. P., 1952- Description: xxiv, 581 p., [8] p. of col. plates: ill.; 24 cm. ISBN: 3764357746 (hbk.: acid-free paper) 0817657746 Notes: Includes bibliographical references and index. Subjects: Nanotechnology. Biotechnology. Combinatorial chemistry. Molecular biology--Automation. Series: Biomethods; vol. 10 LC Classification: TP248.25.N35 M53 1999 Dewey Class No.: 572 21

Milestones in biotechnology: classic papers on genetic engineering / edited by Julian Davies, William S. Reznikoff. Published/Created: Boston: Butterworth-Heinemann, c1992. Related Authors: Davies, Julian

National Symposium on "Perspectives in Biotechnology" (1999: Jodhpur, India) Perspectives in biotechnology: proceedings of National Symposium on "Perspectives in Biotechnology", 1999 / editors, S.M. Reddy, Digamber Rao, Vidyavati. Published/Created: Jodhpur: Scientific Publishers (India), c2001. Related Authors: Reddy, S. M. Rao, Digamber. Vidyavati. Description: viii, 195 p.: ill. (some col.); 25 cm. ISBN: 8172332556 Notes: Includes bibliographical references. Subjects: Microbial biotechnology--Congresses. LC Classification: TP248.27.M53 N38 1999

NATO Advanced Research Workshop on Advances in Animal Cell Technology (1987: Brussels, Belgium) Advanced research on animal cell technology / edited by Alain O.A. Miller. Published/Created: Dordrecht; Boston: Kluwer Academic, c1989. Related Authors: Miller, Alain O. A., 1936- Description: ix, 421 p.: ill.; 25 cm. ISBN: 0792300319 Notes: "Proceedings of the NATO Advanced Research Workshop on Advances in Animal Cell Technology, Brussels, Belgium, September 21-24, 1987"--Verso t.p. Includes bibliographies. Subjects: Animal cell biotechnology--Congresses. Cytology--Congresses. Series: NATO ASI series. Series E, Applied sciences; no. 156. Variant Series: NATO ASI series. Series E, Applied sciences; vol. 156 LC Classification: TP248.27.A53 N37 1987 Dewey Class No.: 660/.6 19

NATO Advanced Research Workshop on Plant Vacuoles: Their Importance in Plant Cell Compartmentation and Their Applications in Biotechnology (1986: Sophia-Antipolis, France) Plant vacuoles: their importance in solute compartmentation in cells and their applications in plant biotechnology / edited by B. Marin. Published/Created: New York: Plenum Press, c1987.

Related Authors: Marin, Bernard P. North Atlantic Treaty Organization. Scientific Affairs Division. Description: xvi, 562 p.: ill., port.; 26 cm. ISBN: 0306426137 Notes: "Proceedings of a NATO Advanced Research Workshop on Plant Vacuoles: Their Importance in Plant Cell Compartmentation and Their Applications in Biotechnology, held July 6-11, 1986, in Sophia-Antipolis, France"--T.p. verso. "Published in cooperation with NATO Scientific Affairs Division." Includes bibliographies and index. Subjects: Plant vacuoles--Congresses. Plant cell compartmentation--Congresses. Plant cell culture--Congresses. Plants--Metabolism--Congresses. Plant biotechnology--Congresses. Biotechnology--congresses. Cell Compartmentation--congresses. Cells, Cultured--congresses. Organoids--metabolism--congresses. Plants--metabolism--congresses. Series: NATO ASI series. Series A, Life sciences; v. 134 LC Classification: QK725 .N37 1986 Dewey Class No.: 581.87/4 19

NATO Advanced Research Workshop on RNA: Biochemistry, and Biotechnology (1998: Poznan, Poland) RNA biochemistry and biotechnology / edited by Jan Barciszewski, Brian F.C. Clark. Published/Created: Dordrecht; Boston: Kluwer, c1999. Related Authors: Barciszewski, Jan. Clark, Brian F. C. (Brian Frederic Carl) Description: xv, 370 p.: ill.; 25 cm. ISBN: 0792358619 (alk. paper) Notes: "Proceedings of NATO Advanced Research Workshop on RNA, Biochemistry, and Biotechnology, held in Poznan, Poland, on October 10-17, 1998"--T.p. verso. Includes bibliographical references and index. Subjects: RNA--Congresses. RNA--Biotechnology--Congresses.

Series: NATO science series. Partnership sub-series 3, High technology; v. 70 Variant Series: NATO ASI series. Series 3, High technology; vol. 70 LC Classification: QP623 .N386 1998 Dewey Class No.: 572.8/8 21

NATO Advanced Research Workshop on Safety Assurance for Environmental Introductions of Genetically-Engineered Organisms (1987: Rome, Italy) Safety assurance for environmental introductions genetically-engineered organisms / edited by Joseph Fiksel, Vincent T. Covello. Published/Created: Berlin; New York: Springer-Verlag, c1988. Related Authors: Fiksel, Joseph R. Covello, Vincent T. North Atlantic Treaty Organization. Scientific Affairs Division. Description: viii, 282 p.: ill.; 25 cm. ISBN: 0387185615 (U.S.) Notes: "Proceedings of the NATO Advanced Research Workshop on Safety Assurance for Environmental Introductions of Genetically-Engineered Organisms held in Rome, Italy, June 6-10, 1987"--T.p. verso. "Published in cooperation with NATO Scientific Affairs Division." Subjects: Biotechnology--Safety measures--Congresses. Genetic engineering--Safety measures--Congresses. Biotechnology--Environmental aspects--Congresses. Series: NATO ASI series. Series G, Ecological sciences; no. 18. Variant Series: NATO ASI series. Series G, Ecological sciences; vol. 18 LC Classification: TP248.14 .N37 1987 Dewey Class No.: 363.17 19

NATO Advanced Research Workshop on the Role of Computational Models and Theories in Biotechnology (1991: San Felíu de Guixols, Spain) Molecular aspects of biotechnology: computational models and theories: [proceedings of the

NATO Advanced Research Workshop on the Role of Computational Models and Theories in Biotechnology, Sant Feliu de Guíxols, Spain, 13-19 June 1991] / edited by J. Bertrán. Published/Created: Dordrecht; Boston: Kluwer Academic Publishers, c1992. Related Authors: Betrán, J. (Juan), 1931- North Atlantic Treaty Organization. Scientific Affairs Division. Description: xiv, 332 p.: ill. (some col.); 25 cm. ISBN: 0792317289 (alk. paper) Notes: Published in cooperation with NATO Scientific Affairs Division. Includes bibliographical references and index. Subjects: Biochemistry--Computer simulation--Congresses. Biotechnology--Computer simulation--Congresses. Series: NATO ASI series. Series C, Mathematical and physical sciences; no. 368. Variant Series: NATO ASI series. Series C, Mathematical and physical sciences; vol. 368 LC Classification: QP517.M3 N38 1991 Dewey Class No.: 574.19/2/0113 20

NATO Advanced Study Institute on Chromatography and Membrane Processes in Biotechnology (1990: São Miguel Island, Azores) Chromatographic and membrane processes in biotechnology / edited by Carlos A. Costa and Joaquim S. Cabral. Published/Created: Dordrecht; Boston: Kluwer Academic Publishers, c1991. Related Authors: Costa, Carlos A. Cabral, Joaquim S. Description: ix, 473 p.: ill.; 25 cm. ISBN: 0792314174 (acid-free) Notes: "Proceedings of the NATO Advanced Study Institute on Chromatography and Membrane Processes in Biotechnology, S. Miguel-Açores, Portugal, July 15-27, 1990"--T.p. verso. Includes bibliographical references and index. Subjects:

Membrane separation--Congresses. Affinity chromatography--Congresses. Biotechnology--Technique--Congresses. Series: NATO ASI series. Series E, Applied sciences; no. 204. Variant Series: NATO ASI series. Series E, Applied sciences, vol. 204. LC Classification: TP248.25.M46 N38 1990 Dewey Class No.: 660/.2842 20

NATO Advanced Study Institute on Plant Cell Biotechnology (1987: Albufeira, Portugal) Plant cell biotechnology / edited by M. Salomé S. Pais, F. Mavituna, J.M. Novais. Published/Created: Berlin; New York: Springer-Verlag, c1988. Related Authors: Pais, Maria Salomé S. Mavituna, Ferda, 1951- Novais, J. M. Description: xx, 500 p.: ill.; 25 cm. ISBN: 0387185569 (U.S.) Notes: "Proceedings of the NATO Advanced Study Institute on Plant Cell Biotechnology held in Albufeira, Algarve, Portugal, March 29-April 10, 1987"--T.p. verso. Includes bibliographies. Subjects: Plant biotechnology--Congresses. Series: NATO ASI series. Series H, Cell biology; vol. 18 LC Classification: TP248.27.P55 N37 1987 Dewey Class No.: 660/.6 19

Nature biotechnology. Published/Created: New York, NY: Nature Pub. Co., [1996- Related Authors: Nature Publishing Company. Description: v.: ill.; 28 cm. Vol. 14, no. 3 (Mar. 1996)- Current Frequency: Monthly, Former Frequency: Monthly, with a special issue in Nov. Continues: Bio/technology (Nature Publishing Company) 0733-222X (DLC) 83644551 (OCoLC)8549852 ISSN: 1087-0156 Cancel/Invalid LCCN: sn 96001682 CODEN: NABIF9 Notes: Title from cover. Supplements accompany some volumes. SERBIB/SERLOC merged record Complements the journal Nature with articles for an audience specialized in biotechnology. Nature 0028-0836 (DLC) 12037118 (OCoLC)1586310 Indexed entirely by: Index medicus 0019-3879 Jan. 1997- Indx'd selectively by: Bibliography of agriculture x 0006-1530 Chemical abstracts 0009-2258 Chemical industry notes 0045-639X Additional Form Avail.: Issued also in microform by University Microfilms, Inc. Also available online via the World Wide Web by subscription. Subjects: Biochemical engineering--Periodicals. Biotechnology--Periodicals. Biotechnology--periodicals. LC Classification: TP248.3 .B557 Dewey Class No.: 660/.6/05 19

Near-infrared applications in biotechnology / edited by Ramesh Raghavachari. Published/Created: New York: M. Dekker, c2001. Related Authors: Raghavachari, Ramesh. Description: xvi, 382 p.: ill.; 24 cm. ISBN: 0824700090 (alk. paper) Notes: Includes bibliographical references and index. Subjects: Near infrared spectroscopy. Biotechnology. Series: Practical spectroscopy; v. 25 LC Classification: TP248.25.N42 N43 2001 Dewey Class No.: 660.6/028/4 21

New developments in biotechnology: ownership of human tissues and cells / Office of Technology Assessment task force, Gladys B. White ... [et al.]. Edition Information: Authorized hardbound ed. Published/Created: Philadelphia: Science Information Resource Center, 1988. Related Authors: White, Gladys B. United States. Congress. Office of Technology Assessment. Description: vii, 168 p.: ill.; 26 cm. ISBN: 0397530021: Notes: Reprint. Originally published: Washington, D.C.: Congress of the U.S., Office of Technology Assessment, [1987]. Includes bibliographies and index. Subjects: Genetic engineering--Law and legislation--United States. Biotechnology industries--Law and legislation--United States. Cell lines. Tissues. Biotechnology. Cell Line. Ownership. LC Classification: KF3827.G4 N49 1988 Dewey Class No.: 343.73/0766606 347.303766606 19

New developments in marine biotechnology / edited by Y. Le Gal and H.O. Halvorson with the editorial assistance of Anne-Marie Lambert. Published/Created: New York: Plenum Press, c1998. Related Authors: Le Gal, Yves. Halvorson, Harlyn O. International Marine Biotechnology Conference (4th: 1997: Sorrento, Italy, etc.) Description: xvi, 343 p.: ill.; 26 cm. ISBN: 0306459078 Notes: "Proceedings of the 4th International Marine Biotechnology Conference, held September 22-29, 1997, in Sorrento, Paestum, Oranto, and Pugnoch

OHOLO Conference (35th: 1990: Elat, Israel) Biologicals from recombinant microorganisms and animal cells: production and recovery: proceedings of the 34th [i.e. 35th] Oholo Conference, Eilat, Israel, 1990 / edited by M.D. White, S. Reuveny, and A. Shafferman. Published/Created: Rehovot, Israel: Balaban Publishers; Weinheim; New York: VCH, c1991. Related Authors: White, M. D. (Moshe D.) Reuveny, S. (Shaul) Shafferman, A. (Avigdor) Description: xiv, 567 p.: ill.; 25 cm. ISBN: 0895739674 (alk. paper) Notes: Includes bibliographical references and indexes. Subjects: Recombinant proteins--Congresses. Microbial biotechnology--Technique--Congresses. Animal cell biotechnology--Technique--Congresses. Biological Products--congresses. Cells, Cultured--congresses. Genetic Engineering--methods--congresses. Recombinant Proteins--isolation & purification congresses. Recombination, Genetic--congresses. LC Classification: TP248.65.P76 O37 1990 Dewey Class No.: 660/.6 20

Oliver, Richard W., 1946- The coming biotech age: the business of bio-materials / Richard W. Oliver. Published/Created: New York: McGraw Hill, c2000. Description: xi, 266 p.: ill., maps; 24 cm. ISBN: 0071350209 (cloth) Notes: Includes index. Subjects: Biotechnology industries--Forecasting. Biotechnology--Forecasting. Bioengineering--Forecasting. Molecular biology--Forecasting. LC Classification: HD9999.B442 O44 2000 Dewey Class No.: 338.4/76606 21

Olson, Steve, 1956- Biotechnology: an industry comes of age / by Steve Olson; for the Academy Industry Program ... [et al.]. Published/Created: Washington, D.C.: National Academy Press, 1986. Related Authors: Academy Industry Program (National Research Council (U.S.)) National Academy of Engineering. Institute of Medicine (U.S.) Description: viii, 120 p., 8 p. of plates: ill. (some col.); 26 cm. ISBN: 0309036313 (pbk.) Notes: Based on a conference held in Washington, D.C., Feb. 27-28, 1985, and sponsored by the Academy Industry Program of the National Academy of Sciences, National Academy of Engineering, and Institute of Medicine. Includes bibliographies and index. Subjects: Biotechnology--Congresses. Genetic

engineering--Congresses. LC Classification: TP248.14 .B56 1986 Dewey Class No.: 660/.6 19

Patents on biotechnological inventions:: the E.C. directive / Gerald Kamstra ... [et al.]. Published/Created: London: Sweet & Maxwell, 2002. Related Authors: Kamstra, Gerald. Description: xvii, 230 p.; 26 cm. Cancelled ISBN: 0752007065 Notes: Includes bibliographical references and index. Subjects: Biotechnology--European Union countries--Patents. Series: Special report (Sweet & Maxwell) Variant Series: Special report LC Classification: KJE2751.B (Biotech...) P+

Penicillium and acremonium / edited by John F. Peberdy. Published/Created: New York: Plenum Press, c1987. Related Authors: Peberdy, John F., 1937- Description: xv, 297 p.: ill.; 24 cm. ISBN: 0306423456 Notes: Includes indexes. Bibliography: p. 285-286. Subjects: Fungi--Biotechnology. Penicillium--Biotechnology. Acremonium--Biotechnology. Series: Biotechnology handbooks; v. 1 LC Classification: TP248.27.F86 P46 1987 Dewey Class No.: 660/.62 19

Phage display of peptides and proteins: a laboratory manual / edited by Brian K. Kay, Jill Winter, John McCafferty. Published/Created: San Diego: Academic Press, 1996. Related Authors: Kay, Brian K. Winter, Jill. McCafferty, John, Dr. Description: xxii, 344 p.: ill. (some col.); 24 cm. ISBN: 0124023800 (alk. paper) Notes: Includes bibliographical references and index. Subjects: Bacteriophages--Laboratory manuals. Microbial biotechnology--Laboratory manuals. Viral proteins--

Laboratory manuals. Affinity chromatography--Laboratory manuals. Peptides--Biotechnology--Laboratory manuals. Proteins--Biotechnology--Laboratory manuals. LC Classification: QR342 .P456 1996 Dewey Class No.: 576/.6482 20

Phage display: a laboratory manual / by Carlos F. Barbas III ... [et al.]. Published/Created: Cold Spring Harbor, NY: Cold Spring Harbor Laboratory Press, c2001. Related Authors: Barbas, Carlos F. Description: 1 v. in various pagings: ill.; 27 cm. ISBN: 0879695463 (cloth: alk. paper) 0879695455 (pbk.: alk. paper) Notes: Includes bibliographical references and index. Subjects: Bacteriophages--Laboratory manuals. Microbial biotechnology--Laboratory manuals. Viral proteins--Laboratory manuals. Affinity chromatography--Laboratory manuals. Peptides--Biotechnology--Laboratory manuals. Proteins--Biotechnology--Laboratory manuals. LC Classification: QR342 .P454 2001 Dewey Class No.: 579.2/6/078 21

Photosynthetic microorganisms in environmental biotechnology / edited by Hiroyuki Kojima, Yuan Kun Lee. Published/Created: New York: Springer, 2001. Projected Pub. Date: 0101 Related Authors: Kojima, Hiroyuki, 1944- Lee, Y. K. (Yuan Kun) Description: p.; cm. ISBN: 9624301360 Notes: Includes bibliographical references. Subjects: Bioremediation. Photosynthetic bacteria. Microalgae. Biodegradation. Algae. Biotechnology. Environmental Pollution--prevention & control. Gram-Negative Oxygenic Photosynthetic Bacteria. LC Classification: TD192.5 .P46 2001

Dewey Class No.: 628.5 21

Physics and chemistry basis of biotechnology / edited by Marcel De Cuyper and Jeff W.M. Bulte. Published/Created: Dordrecht [Netherlands]; Boston: Kluwer Academic Publishers, c2001. Related Authors: Cuyper, Marcel de. Bulte, Jeff W. M. Description: 334 p.: ill.; 25 cm. ISBN: 0792370910 (alk. paper) Notes: Includes bibliographical references and index. Subjects: Biotechnology--Research--Methodology--Congresses. Biochemistry--Congresses. Biophysics--Congresses. Series: Focus on biotechnology; v. 7 LC Classification: TP248.14 .P496 2001 Dewey Class No.: 660.6 21

Pichia protocols / edited by David R. Higgins and James M. Cregg. Published/Created: Totowa, N.J.: Humana Press, c1998. Related Authors: Higgins, David R. Cregg, James M. Description: xi, 270 p.: ill.; 24 cm. ISBN: 0896034216 (alk. paper) Notes: Includes bibliographical references and index. Subjects: Pichia pastoris. Yeast fungi--Biotechnology. Genetic vectors. Pichia--genetics. Series: Methods in molecular biology (Clifton, N.J.); v. 103. Variant Series: Methods in molecular biology; v. 103 LC Classification: TP248.27.Y43 P53 1998 8 Dewey Class No.: 660.6/2 21

Plant and animal cells: process possibilities / editors, C. Webb, and F. Mavituna. Published/Created: Chichester, West Sussex: Ellis Horwood; New York, NY: Halsted Press [distributor], 1987. Related Authors: Webb, Colin. Mavituna, Ferda, 1951- Institution of Chemical Engineers (Great Britain) University of Manchester. Institute of

Science and Technology. Description: 307 p.: ill.; 25 cm. ISBN: 0745801455: Notes: A conference organised by the Institution of Chemical Engineers and the University of Manchester Institute of Science and Technology. Includes bibliographies and index. Subjects: Cell culture--Industrial applications--Congresses. Biotechnology--Congresses. LC Classification: TP248.25.C44 P57 1987 Dewey Class No.: 660/.6 19

Plant biotechnology and development / editor, Peter M. Gresshoff. Published/Created: Boca Raton: CRC Press, c1992. Related Authors: Gresshoff, Peter M., 1948- Description: 171 p.: ill.; 28 cm. ISBN: 0849382610 Notes: Includes bibliographical references and index. Subjects: Plant molecular biology. Plant biotechnology. Plants--Development. Series: A CRC series of current topics in plant molecular biology LC Classification: QK728 .P52 1992 Dewey Class No.: 583/.3220488 20

Plant tissue culture in relation to biotechnology: proceedings of a seminar held in the Royal Irish Academy, 22-23 February, 1982 / edited by A.C. Cassells and J.A. Kavanagh. Published/Created: Dublin: Royal Irish Academy, 1983. Related Authors: Cassells, A. C. Kavanagh, J. A. Royal Irish Academy. Description: 129 p.: ill.; 29 cm. ISBN: 0901714259 (pbk.) Notes: Includes bibliographical references. Subjects: Plant tissue culture--Congresses. Plant micropropagation--Congresses. Plant biotechnology. Plant breeding--Congresses. LC Classification: SB123.6 .P524 1983 Dewey Class No.: 581/.0724 19

Plunkett's biotech & genetics industry almanac / editor and publisher, Jack W. Plunkett. Published/Created: Houston, Tex.: Plunkett Research, c2001. Related Authors: Plunkett, Jack W. Description: viii, 584 p.: ill.; 28 cm. + 1 computer optical disc (4 3/4 in.) ISBN: 1891775189 Notes: "The only comprehensive guide to biotech companies and trends"--Half-title page. Includes index. Subjects: Biotechnology--Directories. Genetics--Directories. Biotechnology industries--Directories. Genetic engineering industry--Directories. LC Classification: TP248.17 .P58 2001 Dewey Class No.: 338.4/36606/025 21

Pollen biotechnology: gene expression & allergen characterization / edited by Shyam S. Mohapatra & R. Bruce Knox. Published/Created: New York: Chapman & Hall, c1996. Related Authors: Mohapatra, Shyam S., 1955- Knox, R. Bruce, 1938- Description: xv, 288 p.: ill.; 24 cm. ISBN: 0412035219 Notes: Includes bibliographical references and index. Subjects: Allergens--Biotechnology. LC Classification: TP248.65.A39 P65 1996 Dewey Class No.: 616.2/02 20

Poly(ethylene glycol) chemistry: biotechnical and biomedical applications / edited by J. Milton Harris. Published/Created: New York: Plenum Press, c1992. Related Authors: Harris, J. Milton. Related Titles: Polyethylene glycol chemistry. Description: xxi, 385 p.: ill.; 24 cm. ISBN: 0306440784 Notes: Includes bibliographical references and index. Subjects: Polyethylene glycol--Biotechnology. Series: Topics in applied chemistry LC Classification: TP248.65.P58 P65 1992

Dewey Class No.: 610/.28 20

Polymer applications for biotechnology: macromolecular separation and identification / David S. Soane, editor. Published/Created: Englewood Cliffs, N.J.: Prentice Hall, c1992. Related Authors: Soane, David S., 1951- Description: xiii, 314 p.: ill.; 24 cm. ISBN: 0136832512 Notes: Includes bibliographical references and index. Subjects: Biomolecules--Separation. Polymers--Biotechnology. Biotechnology--Methodology. Series: Prentice Hall polymer science and engineering. Variant Series: Prentice Hall polymer science and engineering series LC Classification: TP248.25.S47 P65 1992 Dewey Class No.: 660/.6 20

Polysaccharides: structural diversity and functional versatility / edited by Severian Dumitriu. Published/Created: New York: Marcel Dekker, c1998. Related Authors: Dumitriu, Severian, 1939- Description: xiii, 1147 p.: ill.; 26 cm. ISBN: 0824701275 (alk. paper) Notes: Includes bibliographical references and index. Subjects: Polysaccharides--Biotechnology. Polysaccharides--Structure. Microbial polysaccharides. LC Classification: TP248.65.P64 P66 1998 Dewey Class No.: 547/.782 21

Preparative biochemistry & biotechnology: an international journal for rapid communications. Published/Created: New York, NY: M. Dekker, c1996- Description: v.: ill.; 23 cm. Some no. issued together. Vol. 26, no. 1 (Feb. 1996)- Current Frequency: Four no. a year Continues: Preparative biochemistry 0032-7484 (DLC) 72217500 (OCoLC)1589130 ISSN: 1082-6068 Cancel/Invalid LCCN: sn 95004848 CODEN: PBBIF4 Notes: SERBIB/SERLOC merged record Indexed by: ASCA BIOSIS Current advances in plant science 0306-4484 Current contents. Life sciences 0011-3409 International Cancer Research Data Bank (ICRDB) program ISI/BIOMED Journal of abstracts of the All-Union Institute of Scientific and Technical Information of the USSR Science citation index 0036-827X Selected water resources abstracts 0037-136X Indexed entirely by: Index medicus 0019-3879 Feb. 1996- Indx'd selectively by: Bibliography of agriculture 0006-1530 Biological abstracts 0006-3169 Chemical abstracts 0009-2258 Energy research abstracts Mar. 1979- 0160-3604 Excerpta medica International aerospace abstracts 0020-5842 1983- Life sciences collection Subjects: Biochemistry--Technique--Periodicals. Biotechnology--Technique--Periodicals. Biochemistry--methods--periodicals. Biotechnology--methods--periodicals. LC Classification: QH324 .P67 Dewey Class No.: 574.1/92/028

Preparative biotransformations: whole cell and isolated enzymes in organic systems / editor, S.M. Roberts, associate editors, K. Wiggins, G. Casy. Published/Created: Chichester; New York: J. Wiley, c1992- Projected Pub. Date: 1111 Related Authors: Roberts, Stanley M. Wiggins, K. (Karen) Casy, G. (Guy) Description: p. cm. ISBN: 0471929867 Notes: Includes bibliographical references and index. Subjects: Enzymes--Biotechnology. Biotransformation (Metabolism) Organic compounds--Synthesis. Dewey Class No.: 660/.634 20

Protein biotechnology: isolation, characterization, and stabilization /

edited by Felix Franks. Published/Created: Totowa, N.J.: Humana Press, c1993. Related Authors: Franks, Felix. Description: xi, 592 p.: ill.; 24 cm. ISBN: 0896032302 Notes: Includes bibliographical references and index. Subjects: Proteins--Biotechnology. Proteins--analysis. Proteins--physiology. Biotechnology--methods. Series: Biological methods LC Classification: TP248.65.P76 P72 1993 Dewey Class No.: 660/.63 20

Protein C and related anticoagulants / editors, Duane F. Bruley, William N. Drohan. Published/Created: The Woodlands, Tex.: Portfolio Pub. Co.; Houston: Gulf Pub. Co., Book Division, c1990. Related Authors: Bruley, Duane F. Drohan, William. Description: xii, 211 p.: ill.; 24 cm. ISBN: 0943255147: Notes: Includes bibliographical references and index. Subjects: Protein C--Biotechnology. Protein C--Therapeutic use--Testing. Blood coagulation Disorders--therapy. Protein C--isolation & purification. Protein C--therapeutic use. Series: Advances in applied biotechnology series; v. 11 LC Classification: QP93.7.P76 P75 1990 Dewey Class No.: 616.1/57061 20

Protein engineering: approaches to the manipulation of protein folding / edited by Saran A. Narang. Published/Created: Boston: Butterworths, c1990. Related Authors: Narang, Saran A. Description: xix, 262 p.: ill.; 25 cm. ISBN: 0409901164 Notes: Includes bibliographical references and index. Subjects: Proteins--Biotechnology. Series: Biotechnology (Reading, Mass.); 14. Variant Series: Biotechnology; 14 LC Classification: TP248.65.P76 P74 1990

Dewey Class No.: 660/.63 20

Protein expression and purification. Published/Created: San Diego: Academic Press, c1990- Description: v.: ill.; 28 cm. Vol. 1 complete in 2 issues; some issues published in combined form. Vol. 1, no. 1 (Sept. 1990)- Current Frequency: Monthly (except Jan., May, Sept.), <1997- Former Frequency: Bimonthly ISSN: 1046-5928 Cancel/Invalid LCCN: sn 89003262 CODEN: PEXPEJ Notes: Title from cover. Published: Orlando, FL, Apr./June 1991- SERBIB/SERLOC merged record Indexed entirely by: Index medicus 0019-3879 Sept. 1990- Indx'd selectively by: Biological abstracts 0006-3169 1990- Chemical abstracts 0009-2258 1990- Additional Form Avail.: Also available to subscribers via the World Wide Web. Subjects: Proteins--Purification--Periodicals. Proteins--Biotechnology--Periodicals. Gene Expression Regulation--periodicals. Proteins--isolation & purification--periodicals. Proteins--periodicals. LC Classification: QP551 .P695818 Dewey Class No.: 574.19/245 20

Protein immobilization: fundamentals and applications / edited by Richard F. Taylor. Published/Created: New York: M. Dekker, c1991. Related Authors: Taylor, Richard F., 1946- Description: x, 377 p.: ill.; 24 cm. ISBN: 0824782712 (alk. paper) Notes: Includes bibliographical references and index. Subjects: Immobilized proteins. Biotechnology--methods. Proteins. Series: Bioprocess technology; v. 14 LC Classification: TP248.65.I47 P76 1991 Dewey Class No.: 660/.63 20

Protein production by biotechnology / edited by T.J.R. Harris. Published/Created: London; New York: Elsevier Applied Science; New York, NY, USA: Sole distribution in the USA and Canada, Elsevier Science Pub. Co., 1990. Related Authors: Harris, T. J. R. (Tim J. R.) Biological Council. Description: xiii, 243 p.: ill.; 25 cm. ISBN: 1851664017 Notes: Based upon revised and expanded versions of papers presented at the fourth annual biotechnology symposium of the Biological Council held in December 1988 at the Middlesex Hospital Medical School. Includes bibliographical references and index. Subjects: Proteins--Biotechnology--Congresses. Biotechnology--congresses. Protein Engineering--congresses. Series: Elsevier applied biotechnology series LC Classification: TP248.65.P76 P76 1990 Dewey Class No.: 660/.63 20

Protein purification process engineering / edited by Roger G. Harrison. Published/Created: New York: M. Dekker, c1994. Related Authors: Harrison, Roger G., 1944- Description: xiv, 381 p.: ill.; 24 cm. ISBN: 082479009X Notes: Includes bibliographical references and index. Subjects: Proteins--Biotechnology. Proteins--Purification. Series: Bioprocess technology; v. 18 LC Classification: TP248.65.P76 P7654 1994 Dewey Class No.: 660/.63 20

Protein purification: from molecular mechanisms to large-scale processes / Michael R. Ladisch, editor ... [et al.]. Published/Created: Washington, DC: American Chemical Society, 1990. Related Authors: Ladisch, Michael R., 1950- American Chemical Society. Division of Biochemical Technology. American Chemical Society. Meeting (198th: 1989: Miami Beach, Fla.) Description: vii, 280 p.: ill.; 24 cm. ISBN: 0841217904 (alk. paper) Notes: "Developed from a symposium sponsored by the Division of Biochemical Technology at the 198th National Meeting of the American Chemical Society, Miami Beach, Florida, September 10-15, 1989." Includes bibliographical references and indexes. Subjects: Proteins--Purification--Congresses. Proteins--Biotechnology--Congresses. Series: ACS symposium series, 0097-6156; 427 LC Classification: TP248.65.P76 P765 1990 Dewey Class No.: 660/.63 20

Protein-solvent interactions / edited by Roger B. Gregory. Published/Created: New York, N.Y.: M. Dekker, c1995. Related Authors: Gregory, Roger B. Description: xix, 570 p.: ill.; 24 cm. ISBN: 0824792394 Notes: Includes bibliographcial references and index. Subjects: Proteins. Solvation. Hydration. Proteins--Biotechnology. LC Classification: QP551 .P697634 1995 Dewey Class No.: 574.19/245 20

Pseudomonas / edited by Thomas C. Montie. Published/Created: New York: Plenum Press, c1998. Related Authors: Montie, Thomas C. Description: xv, 335 p.: ill.; 24 cm. ISBN: 0306458497 Notes: Includes bibliographical references and index. Subjects: Pseudomonas. Pseudomonas--Biotechnology. Molecular microbiology. Series: Biotechnology handbooks; v. 10 LC Classification: QR82.P78 P77 1998 Dewey Class No.: 579.3/32 21

Pseudomonas: biotransformations, pathogenesis, and evolving biotechnology / edited by Simon Silver

... [et al.]. Published/Created: Washington, D.C.: American Society for Microbiology, c1990. Related Authors: Silver, S. (Simon) American Society for Microbiology. Description: xxiv, 423 p.: ill.; 26 cm. ISBN: 1555810195: Notes: Based on a symposium held in Chicago, Ill., in July, 1989; sponsored by the American Society for Microbiology. Includes bibliographical references and index. Subjects: Pseudomonas--Congresses. Pseudomonas--Biotechnology--Congresses. Pseudomonas--congresses. LC Classification: QR82.P78 P78 1990 Dewey Class No.: 589.9/5 20

Pseudomonas: molecular biology and biotechnology / edited by Enrica Galli, Simon Silver, Bernard Witholt; sponsored by Federation of European Microbiological Societies. Published/Created: Washington, D.C.: American Society for Microbiology, c1992. Related Authors: Galli, Enrica. Silver, S. (Simon) Witholt, Bernard. Federation of European Microbiological Societies. Description: xii, 443 p.: ill.; 26 cm. ISBN: 1555810519 Notes: "Based on the proceedings of a symposium held under the auspices of the Federation of European Microbiological Societies from June 16 to June 20, 1991, in Trieste, Italy"--T.p. verso. Includes bibliographical references and indexes. Subjects: Pseudomonas--Molecular aspects--Congresses. Pseudomonas--Biotechnology--Congresses. Biotechnology--congresses. Pseudomonas--congresses. LC Classification: QR82.P78 P82 1992 Dewey Class No.: 589.9/5 20

Rabinow, Paul. French DNA: trouble in purgatory / Paul Rabinow.

Published/Created: Chicago, IL: University of Chicago Press, 1999. Description: vii, 201 p.; 22 cm. ISBN: 0226701506 (alk. paper) Notes: Includes bibliographical references (p. 183-199). Subjects: Centre d'étude du polymorphisme humain. Millennium Pharmaceuticals, Inc. Human genetics--Government policy--France. Non-insulin-dependent diabetes--Research--United States. Non-insulin-dependent diabetes--Research--France. Human genome. Human gene mapping--France. Biotechnology industries--United States. LC Classification: QH431 .R236 1999 Dewey Class No.: 611/.01816 21

Racek, Jaroslav. Cell-based biosensors / Jaroslav Racek. Published/Created: Lancaster, Pa.: Technomic Pub. Co., c1995. Description: vii, 107 p.: ill.; 23 cm. ISBN: 1566761905 Notes: Includes bibliographical references (p. 95-103) and index. Subjects: Biosensors. Microbial biotechnology. Animal cell biotechnology. Plant cell biotechnology. LC Classification: R857.B54 R33 1995 Dewey Class No.: 574.19/285 20

Rifkin, Jeremy. The biotech century: harnessing the gene and remaking the world / Jeremy Rifkin. Published/Created: New York: Jeremy P. Tarcher/Putnam, c1998. Description: xvi, 271 p.; 24 cm. ISBN: 087477909X (acid-free paper) Notes: Includes bibliographical references (p. [259]-264) and index. Subjects: Biotechnology--Social aspects. Biotechnology--Moral and ethical aspects. Genetic engineering--Social aspects. Genetic engineering--Moral and ethical aspects. LC Classification: TP248.2 .R54 1998 Dewey Class No.: 303.48/3 21 :

Saccharomyces / edited by Michael F. Tuite
and Stephen G. Oliver.
Published/Created: New York: Plenum
Press, c1991. Related Authors: Tuite,
Mick F., 1953- Oliver, S. G. (Stephen
George), 1949- Description: xv, 327 p.:
ill.; 24 cm. ISBN: 0306436345 Notes:
Includes bibliographical references and
index. Subjects: Saccharomyces--
Biotechnology. Saccharomyces. Series:
Biotechnology handbooks; v. 4 LC
Classification: TP248.27.Y43 S22 1991
Dewey Class No.: 660/.62 20

Safety of biological products prepared from
mammalian cell culture: Conference
Centre of the Marcel-Mérieux
Foundation, Les Pensières, Veyrier-du-
Lac, Annecy, France, September 29-
October 1, 1996 / volume editors, Fred
Brown ... [et al.] Published/Created:
Basel; New York: Karger, c1998.
Related Authors: Brown, Fred, 1925-
Fondation Marcel-Mérieux.
International Association of Biological
Standardization. Description: viii, 241
p.: ill.; 24 cm. ISBN: 3805567324
Notes: "Proceedings of a symposium
organized and sponsored by the Marcel-
Mérieux Foundation and the
International Association of Biological
Standardization, and co-sponsored by
the World Health Organization."
Includes bibliographical references.
Subjects: Animal cell biotechnology--
Safety measures--Congresses.
Biological products--Safety measures--
Congresses. Series: Developments in
biological standardization; v. 93 LC
Classification: TP248.27.A53 S255
1998 Dewey Class No.: 660.6 21

Scholz, Carmen, 1963- Polymers from
renewable resources: biopolyesters and
biocatalysts / Carmen Scholz, Richard
A. Gross. Published/Created:

Washington, D.C.: American Chemical
Society, c2000. Related Authors: Gross,
Richard A., 1957- American Chemical
Society. Meeting (216th: 1998: Boston,
Mass.) Description: xii, 356 p.: ill.; 24
cm. ISBN: 0841236461 (alk. paper)
Notes: Includes bibliographical
references and indexes. Subjects:
Polyesters--Congresses. Polymers--
Biodegradation--Congresses. Polymers-
-Biotechnology--Congresses. Series:
ACS symposium series; 764 LC
Classification: TP1180.P6 S36 2000
Dewey Class No.: 668.9 21

Schügerl, K. (Karl) Solvent extraction in
biotechnology: recovery of primary and
secondary metabolites / Karl Schügerl.
Published/Created: Berlin; New York:
Springer-Verlag, c1994. Description:
viii, 213 p.: ill.; 24 cm. ISBN:
0387576940 (New York: alk. paper)
3540576940 (Berlin: alk. paper) Notes:
Includes bibliographical references p.
([198]-207) and index. Subjects:
Extraction (Chemistry) Biotechnology--
Methodology. Metabolites--Separation.
LC Classification: TP248.25.E88 S38
1994 Dewey Class No.: 660/.284248 20

Science Writers Workshop on
Biotechnology and the Human Genome
(1987: Brookhaven National
Laboratory) Biotechnology and the
human genome: innovations and impact
/ edited by Avril D. Woodhead and
Benjamin J. Barnhart; technical editor,
Katherine Vivirito. Published/Created:
New York: Plenum Press, c1988.
Related Authors: Woodhead, Avril D.
Barnhart, Benjamin J. Related Titles:
Human genome. Description: viii, 175
p.: ill.; 26 cm. ISBN: 030642990X
Notes: "Based on the Science Writers
Workshop on Biotechnology and the
Human Genome, held September 14-16,

1987, at Brookhaven National Laboratory, Upton, New York"--T.p. verso. Includes bibliographies and index. Subjects: Human chromosomes--Analysis--Congresses. Human gene mapping--Congresses. Biotechnology--Congresses. Base Sequence--congresses. Biotechnology--congresses. Chromosome Mapping--congresses. Genetic Intervention--congresses. Series: Basic life sciences; v. 46 LC Classification: QH431 .S3778 1987 Dewey Class No.: 573.2/12 19

Space Bioreactor Science Workshop (1985: Houston, Tex.) Space Bioreactor Science Workshop: proceedings of a workshop held at the Nassau Bay Holiday Inn, Houston, Texas, August 22-23, 1985 / Dennis R. Morrison, editor. Published/Created: Washington, D.C.: National Aeronautics and Space Administration, Scientific and Technical Information Division; Springfield, Va.: For sale by National Technical Information Service, 1987. Related Authors: Morrison, Dennis R. United States. National Aeronautics and Space Administration. Description: x, 178 p.: ill.; 28 cm. Notes: Includes bibliographies. Subjects: Bioreactors--Congresses. Space biology--Congresses. Biotechnology--Congresses. Series: NASA conference publication; 2485 LC Classification: TP248.25.B55 S62 1985 Dewey Class No.: 660/.6 20

Spallone, Patricia. Generation games: genetic engineering and the future for our lives / Pat Spallone. Published/Created: Philadelphia: Temple University Press, c1992. Description: viii, 343 p.; 21 cm. ISBN: 087722966X (cloth) 0877229678 (paper) Notes: Includes bibliographical references (p. [321]-324) and index.

Subjects: Genetic engineering. Genetic engineering--Moral and ethical aspects. Biotechnology. Biotechnology--Moral and ethical aspects. Feminist theory. LC Classification: QH442 .S68 1992 Dewey Class No.: 303.48/3 20

Stankiewicz, Rikard. The single cell protein as a technological field / Rikard Stankiewicz. Published/Created: Lund: Research Policy Institute, University of Lund, 1981. Description: 54 p.; 21 cm. ISBN: 9186002120 (pbk.) Notes: Bibliography: p. 53-54. Subjects: Single cell proteins--Biotechnology. Series: Sweden's industrial future Research policy studies. Discussion paper, 0349-1676; no. 142 LC Classification: TP248.65.S56 S72 1981 Dewey Class No.: 660/.63 20

Steinberg, Mark (Mark L.) The Facts on File dictionary of biotechnology and genetic engineering / Mark L. Steinberg, Sharon D. Cosloy. Published/Created: New York: Facts on File, c2001. Related Authors: Cosloy, Sharon D. Facts on File, Inc. Description: x, 228 p.: ill.; 24 cm. ISBN: 0816042748 (alk. paper) 0816042756 (pbk.: alk. paper) Subjects: Biotechnology--Dictionaries. Genetic engineering--Dictionaries. Series: The Facts on File science library LC Classification: TP248.16 .S84 2001 Dewey Class No.: 660.6/03 21

Steinberg, Mark (Mark L.) The Facts on File dictionary of biotechnology and genetic engineering / Mark L. Steinberg, Sharon D. Cosloy; Edmund H. Immergut, series editor. Published/Created: New York: Facts on File, c1994. Related Authors: Cosloy, Sharon D. Description: 197 p.: ill.; 24 cm. ISBN: 0816012504 (acid-free paper) Subjects: Biotechnology--Dictionaries. Genetic engineering--

Dictionaries. LC Classification: TP248.16 .S84 1994 Dewey Class No.: 660/.6/03 20

Stereoselective biocatalysis / edited by Ramesh N. Patel. Published/Created: New York: M. Dekker, c2000. Related Authors: Patel, Ramesh N., 1942- Description: xiii, 932 p.: ill.; 26 cm. ISBN: 0824782828 (acid-free paper) Notes: Includes bibliographical references and index. Subjects: Enzymes--Biotechnology. Stereoisomers--Synthesis. LC Classification: TP248.65.E59 S88 2000 Dewey Class No.: 660.6

Strategies for protein purification and characterization: a laboratory course manual / Daniel R. Marshak ... [et al.]. Published/Created: Plainview, N.Y.: Cold Spring Harbor Laboratory Press, 1996. Related Authors: Marshak, Daniel R. Description: xviii, 396 p.: ill.; 27 cm. ISBN: 0879694491 (cloth) 0879693851 (comb bound) Contents: Purification of calmodulin -- Purification of transcription factor AP-1 from HeLA cells -- Purification of a recombinant protein overproduced in Escherichia coli -- Solubilization and purification of the rat liver insulin receptor. Notes: Includes bibliographical references (p. 385-86) and indexes. Subjects: Proteins--Purification--Laboratory manuals. Proteins--Biotechnology. Protein engineering. Proteins--isolation & purification--laboratory manuals. LC Classification: QP551 .S855 1996 Dewey Class No.: 574.19/245 20

Straus, Joseph. Deposit and release of biological material for the purposes of patent procedure / Joseph Straus in cooperation with Rainer Moufang; English translation by Anthony Rich in cooperation with the authors. Published/Created: Baden-Baden: Nomos Verlagsgesellschaft, 1990. Related Authors: Moufang, Rainer. Description: 173 p.; 23 cm. ISBN: 3789020230 Notes: Translation of: Hinterlegung und Freigabe von biologischem Material für Patentierungszwecke. Includes bibliographical references (p. 169-173). Subjects: Biotechnology--Patents. LC Classification: K1519.B54 S7713 1990 Dewey Class No.: 346.04/86 342.6486 20

Supramolecular design for biological applications / Nobuhiko Yui, editor. Published/Created: Boca Raton, FL: CRC Press, 2002. Projected Pub. Date: 0201 Related Authors: Yui, Nobuhiko. Description: p. cm. ISBN: 0849309654 (alk. paper) Notes: Includes bibliographical references and index. Subjects: Supramolecular chemistry. Biochemistry. Biotechnology. LC Classification: QP801.P64 S87 2002 Dewey Class No.: 572.8 21

Synthetic biodegradable polymer scaffolds / Anthony Atala, David J. Mooney, editors; Joseph P. Vacanti, Robert Langer, associate editors. Published/Created: Boston: Birkhäuser, c1997. Related Authors: Atala, Anthony, 1958- Mooney, David J., 1964- Description: xii, 258 p.: ill.; 24 cm. ISBN: 0817639195 (acid-free paper) 3764339195 (acid-free paper) Notes: Includes bibliographical references and index. Subjects: Polymers in medicine--Biodegradation. Tissue culture. Animal cell biotechnology. Series: Tissue engineering LC Classification: R857.P6 S957 1997 Dewey Class No.: 617.9/5

21

Technical/final report to Saskatchewan Agriculture Development Fund: genetics and biotechnology. Published/Created: [Regina, Sask.]: Saskatchewan Agriculture Development Fund, [1987] Related Authors: Saskatchewan. Agriculture Development Fund. Related Titles: Genetics and biotechnology. Description: 1 v. (various pagings): ill.; 28 cm. Notes: "R-86-05-0075." Includes bibliographical references. Subjects: Crops--Genetic engineering. Crops--Saskatchewan--Genetic engineering. Plant biotechnology--Saskatchewan. LC Classification: SB123.57 .T43 1987

The Bio-revolution: cornucopia or Pandora's box? / edited by Peter Wheale and Ruth McNally. Published/Created: London; Winchester, Mass.: Pluto Press, 1990. Related Authors: Wheale, Peter. McNally, Ruth M. Athene Trust. Athene Trust International Conference (1988: London, England) Description: xviii, 300 p.; 23 cm. ISBN: 0745303374: 0745303382 (pbk.): Notes: Based upon the papers presented at the Athene Trust International Conference held in London in September 1988. Includes bibliographical references and indexes. Subjects: Genetic engineering--Congresses. Biotechnology--Congresses. Series: Genetic engineering series LC Classification: TP248.6 .B53 1990 Dewey Class No.: 660/.65 20

The Genetic revolution: scientific prospects and public perceptions / edited by Bernard D. Davis. Published/Created: Baltimore: Johns Hopkins University Press, c1991. Related Authors: Davis, Bernard D., 1916- American Academy of Arts and Sciences. Description: xvi,

295 p.: ill.; 24 cm. ISBN: 0801842352 (alk. paper) 0801842395 (pbk.) Notes: Based on conference proceedings sponsored by the American Academy of Arts and Sciences. Includes bibliographical references (p. [283]-284) and index. Subjects: Genetic engineering. Biotechnology. LC Classification: QH442 .G463 1991 Dewey Class No.: 660/.65 20

The Impact of biotechnology on the environment: a literature review of environmental applications and impacts of biotechnology. Published/Created: Luxembourg: Office for Official Publication of the European Communities; Washington, DC: European Community Information Service [distributor], 1987. Related Authors: European Foundation for the Improvement of Living and Working Conditions. Description: iv, 68 p.; 30 cm. ISBN: 9282575292 Notes: "European Foundation for the Improvement of Living and Working Conditions." Includes bibliographies. Subjects: Biotechnology--Social aspects--Congresses. Biotechnology--Environmental aspects--Congresses. Agricultural biotechnology--International cooperation Congresses. Series: Research report (European Foundation for the Improvement of Living and Working Conditions) Variant Series: Research report LC Classification: TP248.14 .I45 1987 Dewey Class No.: 660/.6 20

The Impact of chemistry on biotechnology: multidisciplinary discussions / Marshall Phillips, editor ... [et al.]; developed from a symposium sponsored by the Biotechnology Secretariat at the 192nd Meeting of the American Chemical Society, Anaheim, California,

September 7-12, 1986. Published/Created: Washington, DC: The Society, 1988. Related Authors: Phillips, Marshall. American Chemical Society. Biotechnology Secretariat. American Chemical Society. Meeting (192nd: 1986: Anaheim, Calif.) Description: xiii, 415 p.: ill.; 24 cm. ISBN: 0841214468 Notes: Includes bibliographies and indexes. Subjects: Biotechnology--Congresses. Chemistry, Technical--Congresses. Series: ACS symposium series; 362 LC Classification: TP248.14 .I46 1988 Dewey Class No.: 660/.6 19

Tombs, M. P. An introduction to polysaccharide biotechnology / Michael P. Tombs and Stephen E. Harding. Published/Created: London; Bristol, PA: Taylor & Francis, c1998. Related Authors: Harding, S. E. (Stephen E.) Description: viii, 183 p.: ill.; 26 cm. ISBN: 074840516X Notes: Includes bibliographical references and index. Subjects: Polysaccharides--Biotechnology. Polysaccharides--Metabolism. Glucosidases--Biotechnology. LC Classification: TP248.65.P64 T66 1998 Dewey Class No.: 660.6/3 21

Tombs, M. P. An introduction to polysaccharide biotechnology / Michael P. Tombs and Stephen E. Harding. Published/Created: London; Bristol, PA: Taylor & Francis, c1998. Related Authors: Harding, S. E. (Stephen E.) Description: viii, 183 p.: ill.; 26 cm. ISBN: 074840516X Notes: Includes bibliographical references and index. Subjects: Polysaccharides--Biotechnology. Polysaccharides--Metabolism. Glucosidases--Biotechnology. Polysaccharides. Biotechnology. LC Classification:

Acquisition in Process

Treatment of lignocellulosics with white rot fungi: proceedings of a workshop held in Braunschweig (Federal Republic of Germany) from 21 to 23 October 1986 under the auspices of COST ... / edited by F. Zadrazil and P. Reiniger. Published/Created: London; New York: Elsevier Applied Science: Sole distributor in the USA and Canada, Elsevier Science Pub. Co., c1988. Related Authors: Zadrazil, F. Reiniger, P. European Cooperation in Scientific and Technical Research (Organization) Commission of the European Communities. Bundesforschungsanstalt für Landwirtschaft (Germany) Description: vi, 122 p.: ill.; 25 cm. ISBN: 1851662413 Notes: "Organized with the support of the Commission of the European Communities by the Federal Research Centre for Agriculture (FAL), Braunschweig"--Opp. t.p. Includes bibliographical references. Subjects: Lignocellulose--Biotechnology--Congresses. Lignocellulose--Biodegradation--Congresses. Fungi--Biotechnology--Congresses. Series: EUR (Series); 11252. Variant Series: EUR; 11252 LC Classification: TP248.65.C45 T73 1988

Trends in biotechnology. Published/Created: Amsterdam, Netherlands: Elsevier Science Publishers B.V. (Biomedical Division), c1983- Description: v.: ill.; 27 cm. Vol. 1 (1983)- Ceased with vol. 18 (2000). Cf. Letter from vendor. Current Frequency: Annual ISSN: 0167-9430 Cancel/Invalid LCCN: sn 85019817 Notes: Published: Cambridge, UK: Elsevier Trends Journals, <1995- SERBIB/SERLOC merged record Compilation of articles reprinted from the regular bimonthly ed. Subjects: Biotechnology--Periodicals. Biochemistry--methods--periodicals. Biomedical Engineering--trends--periodicals. Technology--periodicals. LC Classification: TP248.13 .T743 Dewey Class No.: 660/.6 19

Trends in plant tissue culture and biotechnology / editor, L.K. Pareek; associate editor, P.L. Swarnkar. Published/Created: Bikaner: Agro Botanical Publishers, 1997. Related Authors: Pareek, L. K. Swarnkar, P. L. Description: 334 p.: ill.; 25 cm. ISBN: 8185031924 Summary: Papers presented at a symposium held in 1994 at Jaipur. Notes: Includes bibliographical references (p. 321-334). Subjects: Plant tissue culture--Congresses. Plant biotechnology--Congresses. LC Classification: QK725 .T74 1997 Dewey Class No.: 571.5/382 21

U.S.-Japan Meeting on Aquaculture (23rd: 1994: Mie-ken, Japan) Biological control and improvement of salmon and advanced concept of the technology for aquaculture: proceedings of the 23rd Japan-U.S.A. Joint Meeting on Aquaculture in Mie, Japan, November 17-18, 1994, followed by its Satellite Symposium in Niigata, November 21, 1994 / edited by Masanori Azeta ... [et al.]. Published/Created: [Tamaki, Watarai, Mie, Japan]: National Research Institute of Aquaculture (NRIA); [Niigata, Japan]: Fisheries Agency, c1996. Related Authors: Azeta, Masanori. Yōshoku Kenkyūjo (Japan) Japan. Suisanchō. Description: vi, 121 p.: ill., maps; 28 cm. Notes: "March 1996." Includes bibliographical references.

United States. Congress. Senate. Committee on Commerce, Science, and Transportation. National Oceanic and Atmospheric Administration Authorization: hearing before the Committee on Commerce, Science, and Transportation, United States Senate, One Hundred Third Congress, first session, June 22, 1993. Published/Created: Washington: U.S. G.P.O.: For sale by the U.S. G.P.O., Supt. of Docs., Congressional Sales Office, 1994. Description: iii, 52 p.; 24 cm. ISBN: 0160435706 Notes: Distributed to some depository libraries in microfiche. Shipping list no.: 94-0045-P. Subjects: United States. National Oceanic and Atmospheric Administration--Appropriations and expenditures. Marine biotechnology--Research--United States. Oceanography--Research--United States. Series: United States. Congress. Senate. S. hrg.; 103-398. Variant Series: S. hrg.; 103-398 LC Classification: KF26 .C69 1993t

United States. Congress. Senate. Committee on Environment and Public Works. Subcommittee on Hazardous Wastes and Toxic Substances. Federal oversight of biotechnology: hearing before the Subcommittee on Hazardous Wastes and Toxic Substances of the Committee on Environment and Public Works, United States Senate, One Hundredth Congress, first session, November 5, 1987. Published/Created: Washington: U.S. G.P.O.: For sale by the Supt. of Docs., Congressional Sales Office, U.S. G.P.O., 1988. Description: iii, 174 p.: ill.; 24 cm. Notes: Distributed to some depository libraries in microfiche. Shipping list no.: 88-127-P. Item 1045-A, 1045-B (microfiche) Bibliography: p. 119. Subjects: Biotechnology industries-

-Government policy--United States. Series: United States. Congress. Senate. S. hrg.; 100-441. Variant Series: S. hrg.; 100-441 LC Classification: KF26 .E65 1987a Dewey Class No.: 660/.6/0973 19

United States. Congress. Senate. Committee on Environment and Public Works. Subcommittee on Toxic Substances and Environmental Oversight. Releasing genetically engineered organisms into the environment: hearing before the Subcommittee on Toxic Substances and Environmental Oversight of the Committee on Environment and Public Works, United States Senate, Ninety-ninth Congress, second session, May 16, 1986. Published/Created: Washington: U.S. G.P.O.: For sale by the Supt. of Docs., Congressional Sales Office, U.S. G.P.O., 1986. Description: iii, 106 p.; 24 cm. Notes: Distributed to some depository libraries in microfiche. Shipping list no.: 86-756-P. Item 1045-A, 1045-B (microfiche) Subjects: Genetic engineering--Environmental aspects--United States. Biotechnology--Environmental aspects--United States. Series: United States. Congress. Senate. S. hrg.; 99-740. Variant Series: S. hrg.; 99-740 LC Classification: KF26 .E678 1986 Dewey Class No.: 363.7 19

Weizer, William P. Biotechnology applications in chemicals / prepared by William Weizer. Published/Created: Cleveland, Ohio (11001 Cedar Ave., Cleveland 44106): Predicasts, 1984. Description: v, 108 leaves; 30 cm. Notes: "332." Subjects: Chemical industry--United States. Biotechnology industries--United States. Market surveys--United States. LC Classification: HD9651.5 .W45 1984 Dewey Class No.: 381/.456208/0973 19

Wells, Donna Koren, 1951- Biotechnology / Donna Wells. Published/Created: Tarrytown, N.Y.: Benchmark Books, c1996. Description: 63 p.: ill. (some col.); 26 cm. ISBN: 076140046X Summary: Describes the history and applications of biotechnology, focusing on how scientists develop new organisms by altering the genetic makeup of living things. Notes: Includes bibliographical references (p. 61) and index. Subjects: Biotechnology--Juvenile literature. Genetic engineering--Juvenile literature. Biotechnology. Genetic engineering. Series: Inventors & inventions LC Classification: TP248.218 .W45 1996 Dewey Class No.: 660/.65 20

Wheelwright, Scott M. Protein purification: design and scale up of downstream processing / Scott M. Wheelwright. Published/Created: Munich; New York: Hanser Publishers; New York: Distributed in the U.S.A. and in Canada by Oxford University Press, c1991. Description: xvii, 228 p.: ill.; 24 cm. ISBN: 3446157034 (Hanser) 0195209133 (Oxford University Press) Notes: Includes bibliographical references and index. Subjects: Proteins--Biotechnology. Proteins--Purification. LC Classification: TP248.65.P76 W44 1991 Dewey Class No.: 660/.63 20

Xylans and xylanases: proceedings of an international symposium, Wageningen, The Netherlands, December 8-11, 1991 / edited by J. Visser ... [et al.]. Published/Created: Amsterdam: New York: Elsevier, 1992. Related Authors: Visser, J. Description: xv, 576 p.: ill.; 25 cm. ISBN: 0444894772 (alk. paper) Notes: Includes bibliographical references. Subjects: Xylans--

Congresses. Xylanases--Congresses.
Xylanases--Biotechnology--Congresses.
Series: Progress in biotechnology; 7 LC
Classification: QP702.X87 X95 1992
Dewey Class No.: 574.19/248 20

YAC libraries: a user's guide / edited by
David L. Nelson, Bernard H.
Brownstein. Published/Created: New
York: W.H. Freeman and Co., c1994.
Related Authors: Nelson, David L.,
1956- Brownstein, Bernard H.
Description: xx, 211 p.: ill.; 23 cm.
ISBN: 0716770148 Notes: Includes
bibliographical references and index.
Subjects: Yeast--Biotechnology. Yeast
fungi--Genetic engineering.
Chromosomes--Catalogs. Series:
UWBC biotechnical resource series LC
Classification: TP248.27.Y43 Y33 1994
Dewey Class No.: 589.2/330415/0724
20

Yeast biotechnology / edited by D.R. Berry,
I. Russell, G.G. Stewart.
Published/Created: London; Boston:
Allen & Unwin, 1987. Related Authors:
Berry, David R. (David Richard)
Russell, Inge, 1947- Stewart, Graham
G., 1942- Description: xx, 539 p.: ill.;
24 cm. ISBN: 0045740429 (alk. paper)
Notes: Includes bibliographies and
index. Subjects: Yeast fungi--
Biotechnology. Saccharomyces
cerevisiae--Biotechnology. LC
Classification: TP248.27.Y43 Y43 1987
Dewey Class No.: 660/.62 19

Yeast hybrid technologies / edited by Li
Zhu, Gregory J. Hannon.
Published/Created: Natick, MA: Eaton
Pub., c2000. Related Authors: Zhu, Li,
1949- Hannon, Gregory J., 1964-
Description: xvi, 379 p.: ill.; 26 cm.
ISBN: 1881299155 Notes:
"BioTechniques books." Includes

bibliographical references and index.
Subjects: Protein binding--Laboratory
manuals. Yeast fungi--Biotechnology--
Laboratory manuals. Transcription
factors--Laboratory manuals. LC
Classification: QP551 .Y397 2000
Dewey Class No.: 572/.6 21

Yeast protocols: methods in cell and
molecular biology / edited by Ivor H.
Evans. Published/Created: Totowa,
N.J.: Humana Press, c1996. Related
Authors: Evans, Ivor H. (Ivor Howell)
Description: xiii, 433 p.: ill.; 23 cm.
ISBN: 0896033198 (acid-free paper)
Notes: Includes bibliographical
references and index. Subjects: Yeast--
Biotechnology. Yeast fungi--Research--
Methodology. Yeasts--physiology.
Yeasts--isolation & purification. Yeasts-
-cytology. Series: Methods in molecular
biology (Clifton, N.J.); v. 53. Variant
Series: Methods in molecular biology;
v. 53 LC Classification: TP248.27.Y43
Y416 1996 Dewey Class No.:
589.2/330487 20

Yeast: biotechnology and biocatalysis /
edited by Hubert Verachtert, René De
Mot. Published/Created: New York: M.
Dekker, c1990. Related Authors:
Verachtert, Hubert, 1932- Mot, René de,
1957- Description: xv, 522 p.: ill.; 24
cm. ISBN: 0824781422 (alk. paper)
Notes: Includes bibliographical
references. Subjects: Yeast fungi--
Biotechnology. Microbial enzymes.
Series: Bioprocess technology; v. 5 LC
Classification: TP248.27.Y43 Y433
1990 Dewey Class No.: 660/.62 20

Yeasts / edited by B.E. Kirsop and C.P.
Kurtzman in collaboration with T.
Nakase and D. Yarrow.
Published/Created: Cambridge
[England]; New York: Cambridge

University Press, 1988. Related
Authors: Kirsop, B. E. Kurtzman, C. P.
Nakase, T. Yarrow, D. Description: xiii,
234 p.: ill.; 24 cm. ISBN: 0521352274
Notes: Includes index. Bibliography: p.
214-223. Subjects: Yeast fungi--
Biotechnology. Series: Living resources
for biotechnology LC Classification:
TP248.27.Y43 Y42 1988 Dewey Class
No.: 660/.62 19

BUSINESS

Acharya, Rohini. The emergence and growth of biotechnology: experiences in industralised and developing countries / Rohini Acharya. Published/Created: Cheltenham, UK; Northampton, MA: E. Elgar, c1999. Description: xiii, 138 p.: ill.; 24 cm. ISBN: 1858985234 Notes: Includes bibliographical references (p. 125-132) and index. Subjects: Biotechnology industries. Pharmaceutical industry. Biotechnology. Biological diversity conservation. Series: New horizons in the economics of innovation LC Classification: HD9999.B442 A25 1999 Dewey Class No.: 338.4/76606 21

Athene Trust Conference on Biotechnology (1988: London, England) The Athene Trust Conference on Biotechnology: London, UK, October 1988. Published/Created: Washington, D.C. (655 15th St., N.W., Washington 20005): Mongoven, Biscoe & Duchin, c1988. Related Authors: Athene Trust. Description: 43 leaves; 29 cm. Subjects: Biotechnology--Moral and ethical aspects--Congresses. LC Classification: TP248.14 .A84 1988 Dewey Class No.: 179/.1 20

Avramovic, Mila, 1954- An affordable development?: biotechnology, economics, and the implications for the third world / Mila Avramovic. Published/Created: London; Atlantic Highlands, N.J.: Zed Books, 1996. Description: xiv, 210 p.; 23 cm. ISBN: 1856493334 (hb) 1856493342 (pb) Notes: Includes bibliographical references (p. [191]-203) and index. Subjects: Biotechnology--Developing countries. Biotechnology industries--Developing countries. LC Classification: TP248.195.D44 A96 1996 Dewey Class No.: 338.4/76606 20

Biotechnology in Europe and Latin America: prospects for co-operation / edited by Bernardo Sorj, Mark Cantley, Karl Simpson. Published/Created: Dordrecht; Boston: Kluwer Academic for the Commission of the European Communities; Norwell, MA, U.S.A.: Distributed in the U.S.A. and Canada by Kluwer Academic, c1989. Related Authors: Sorj, Bernardo. Cantley, M. F. (Mark F.) Simpson, Karl. Seminar on Biotechnology in Europe and Latin America (1987: Brussels, Belgium) Description: xx, 223 p.: ill.; 25 cm. ISBN: 0792302788 (U.S.: alk. paper) Notes: Includes papers presented at SOBELA, a Seminar on Biotechnology in Europe and Latin America, held Brussels, Apr. 1987. Subjects:

Biotechnology--Europe--International cooperation. Biotechnology--Latin America--International cooperation. LC Classification: TP248.195.E8 B56 1989 Dewey Class No.: 338.9406 20

Biotechnology in future society: scenarios and options for Europe / edited by Edward Yoxen, Vittorio Di Martino. Published/Created: Luxembourg: Office for Official Publications of the European Communities; Brookfield, Vt., USA: Gower Pub. Co., c1989. Related Authors: Yoxen, Edward. Di Martino, Vittorio. European Foundation for the Improvement of Living and Working Conditions. Description: xiv, 147 p.; 23 cm. ISBN: 1855210169 (U.S.) Notes: At head of European Foundation for the Improvement of Living and Working Conditions. "Catalogue number SY-52-88-598-EN-C"--T.p. verso. Includes bibliographical references (p. 140-147). Subjects: Biotechnology--Europe--Forecasting. LC Classification: TP248.195.E8 B57 1989 Dewey Class No.: 660/.6/09401 20

Biotechnology in Latin America: politics, impacts, and risks / edited by N. Patrick Peritore and Ana Karina Galve-Peritore. Published/Created: Wilmington, Del.: SR Books, 1995. Related Authors: Peritore, N. Patrick. Galve-Peritore, Ana Karina, 1962- Description: xxiv, 229 p.: ill.; 24 cm. ISBN: 0842025561 (cloth: alk. paper) 084202557X (pbk.: alk. paper) Notes: Includes bibliographical references (p. 211-229). Subjects: Biotechnology--Latin America. Series: Latin American silhouettes LC Classification: TP248.195.L29 B54 1995 Dewey Class No.: 660.6/098 20

Biotechnology in society: private initiatives and public oversight / Joseph G. Perpich, guest editor. Published/Created: New York: Pergamon, c1986. Related Authors: Perpich, Joseph G. Related Titles: [Technology in society. Description: xii, 223 p.: ill.; 26 cm. ISBN: 0080331688: Notes: "Published in part as special issues of Technology in society, an international journal, George Bugliarello, A. George Schillinger ... editors." Includes bibliographies. Subjects: Biotechnology--Social aspects. LC Classification: TP248.2 .B557 1986 Dewey Class No.: 303.4/83 19

Biotechnology in the 1990s: taking stock / Cort Wrotnowski, project analyst. Published/Created: Norwalk, CT: Business Communications Co., c1987. Related Authors: Wrotnowski, Cort. Business Communications Co. Description: xxxiii, 385 p.; 29 cm. ISBN: 0893368865 Notes: "March 1992"--T.p. verso. Subjects: Biotechnology industries--United States. Biotechnology--United States. Market surveys--United States. Series: Business opportunity report; C-143 LC Classification: HD9999.B443 U6155 1987

Biotechnology Industry Organization. Membership directory / BIO. Published/Created: Washington, DC: Biotechnology Industry Organization, 1994- Related Authors: Institute for Biotechnology Information (North Carolina Biotechnology Center) Description: v.; 28 cm. 1994-95- Current Frequency: Annual Continues: BIO directory 1075-6434 (DLC) 94640772 (OCoLC)30093835 ISSN: 1082-3751 Notes: Vols. for 1994-95-

prepared and designed by the Institute for Biotechnology Information of the North Carolina Biotechnology Center. SERBIB/SERLOC merged record Subjects: Biotechnology Industry Organization--Directories. Biotechnology industries--Directories. Biotechnology industries--United States--Directories. LC Classification: HD9999.B44 B53 Dewey Class No.: 338 12

Biotechnology Law Institute. Annual Biotechnology Law Institute: [proceedings]. Published/Created: New York: Law & Business/Harcourt Brace Jovanovich, c1985- Description: v.; 22 cm. 2nd (1985)- Current Frequency: Annual Notes: Materials distributed at the program. SERBIB/SERLOC merged record Subjects: Biotechnology industries--Law and legislation--United States. Biotechnology--United States--Patents. Industrial microbiology--United States--Patents. LC Classification: KF1893.B56 B56 Dewey Class No.: 346.7304/86 347.306486 19

Biotechnology Patent Conference. Biotechnology Patent Conference workbook. Published/Created: Rockville, Md.: American Type Culture Collection, Related Authors: American Type Culture Collection. Description: v.: ill.; 28 cm. Current Frequency: Annual Notes: Description based on: 1984. SERBIB/SERLOC merged record Subjects: Biotechnology--United States--Patents--Congresses. Microorganisms--Patents--Congresses. LC Classification: KF3133.B5 B56 Dewey Class No.: 346.7304/86 347.306486 20

Biotechnology patents: a business manager's legal guide. Published/Created: Washington, D.C.: Special Project Unit

of the Bureau of National Affairs, c1989. Related Authors: Bureau of National Affairs (Washington, D.C.) Description: 1 v. (various pagings); 28 cm. ISBN: 1558711333: Notes: Cover title. Subjects: Biotechnology--United States--Patents. Series: The BNA special report series on biotechnology; special rept. #1 LC Classification: KF3133.B56 B564 1989 Dewey Class No.: 346.7304/86 347.306486 20

Biotechnology policy development / prepared for the Ontario Ministry of the Environment, Project Number 265RR, by the Canadian Environmental Law Research Foundation, June 1988. Published/Created: Toronto, Ont.: Canadian Institute for Environmental Law & Policy, [1988] Related Authors: Ontario. Ministry of the Environment. Canadian Environmental Law Research Foundation. Description: 2 v.; 29 cm. ISBN: 0772942404 (v. 1) 0772942412 (v. 2) Cancelled ISBN: 0772742390 (set) Notes: Includes bibliographical references. Subjects: Biotechnology industries--Law and legislation--Canada. Biotechnology--Environmental aspects--Canada. LC Classification: KE1858.B56 B56 1988 Dewey Class No.: 343.71/0786606 347.103786606 20

Biotechnology, 1982 update. Published/Created: [Washington, D.C.?]: U.S. Dept. of Commerce, and Trademark Office: [Supt. of Docs., U.S. G.P.O., distributor], 1983. Related Authors: United States. Patent and Trademark Office. Office of Technology Assessment and Forecast. Description: iii, 297 p.: ill.; 29 cm. Notes: "A 1983 publication from the Office of Technology Assessment and Forecast." "First published: July, 1982. Reprinted with additional data,

September 1983--P. ii." Distributed to depository libraries in microfiche. Item 256-B (microfiche) Subjects: Biotechnology--Patents. Series: Patent profiles LC Classification: TP248.2 .P38 1983 Dewey Class No.: 660/.6/0272 19

Biotechnology, a publication / from the Office of Technology Assessment & Forecast. Published/Created: [Washington, D.C.?]: U.S. Dept. of Commerce, Patent and Trademark Office: [Supt. of Docs., U.S. G.P.O., distributor], 1982. Related Authors: United States. Patent and Trademark Office. Office of Technology Assessment and Forecast. Description: iii, 253 p.; 29 cm. Notes: July 1982." S/N 003-004-00587-4 Subjects: Biotechnology--United States--Patents. Series: Patent profiles LC Classification: TP248.2 .B547 1982 Dewey Class No.: 660/.6/0272 19

Biotechnology, commercialization, and economic aspects: January 1991 - January 1994 / Kim Guenther ... [et al.]. Published/Created: Beltsville, Md.: National Agricultural Library, [1994] Related Authors: Guenther, Kim. National Agricultural Library (U.S.) Description: 41 p.; 28 cm. Notes: "209 citations in English from AGRICOLA." "Updates QB 92-60." Shipping list no.: 94-0151-P. "April 1994." Includes indexes. Subjects: Biotechnology-- Economic aspects--Bibliography. Genetic engineering--Economic aspects- -Bibliography. Series: Quick bibliography series; 94-20. Variant Series: Quick bibliography series, 1052- 5378; QB 94-20 LC Classification: Z7914.B33 B56 1994 TP248.2

Biotechnology, patents, and morality / edited by Sigrid Sterckx. Edition Information: 2nd ed. Published/Created: Aldershot, England; Burlington, VT: Ashgate, c2000. Related Authors: Sterckx, Sigrid. Description: xxiv, 382 p.: ill.; 23 cm. ISBN: 0754611442 Notes: Includes bibliographical references. Subjects: Biotechnology--Patents--Congresses. Biotechnology--Patents--Moral and ethical aspects Congresses. LC Classification: TP248.2 .B567 2000 Dewey Class No.: 174/.957 21

Biotechnology, patents, and morality / edited by Sigrid Sterckx. Published/Created: Aldershot, Hants., England; Brookfield, Vt.: Ashgate, c1997. Related Authors: Sterckx, Sigrid. Description: xxi, 324 p.: ill.; 23 cm. ISBN: 1840141581 Notes: Includes bibliographical references. Subjects: Biotechnology--Congresses. Biotechnology--Patents--Congresses. Biotechnology--Moral and ethical aspects--Congresses. Biotechnology-- Patents--Moral and ethical aspects Congresses. LC Classification: TP248.2 .B567 1997 Dewey Class No.: 174/.957 21

Biotechnology: assessing social impacts and policy implications / edited by David J. Webber; prepared under the auspices of the Policy Studies Organization. Published/Created: New York: Greenwood Press, 1990. Related Authors: Webber, David J., 1951- Policy Studies Organization. Description: xiv, 239 p.; 25 cm. ISBN: 0313274541 (lib. bdg.: alk. paper) Notes: Includes bibliographical references (p. [211]-230) and index. Subjects: Biotechnology industries-- Social aspects. Series: Contributions in

political science, 0147-1066; no. 260
LC Classification: HD9999.B442 B567
1990 Dewey Class No.: 338.4/76606 20

Biotechnology: business, law, and
regulation: ALI-ABA course of study
materials: November 18-19, 1993, San
Francisco, California / cosponsored by
California Continuing Education of the
Bar. Published/Created: Philadelphia,
PA (4025 Chestnut St., Philadelphia
19104-3099): American Law Institute-
American Bar Association Committee
on Continuing Professional Education,
c1993. Related Authors: California
Continuing Education of the Bar.
Description: xii, 226 p.: ill.; 27 cm.
Notes: "C886." Subjects: Biotechnology
industries--Law and legislation--United
States. LC Classification: KF1893.B56
B555 1993 Dewey Class No.:
346.7303/8 347.30638 20

Biotechnology: economic and social aspects:
issues for developing countries / edited
by E.J. DaSilva, C. Ratledge, and A.
Sasson. Published/Created: Cambridge
[England]; New York, NY, USA:
Cambridge University Press, 1992.
Related Authors: DaSilva, E. J.
Ratledge, Colin. Sasson, Albert.
Description: xiii, 388 p.: ill.; 24 cm.
ISBN: 0521384737 (hardback) Notes:
"Published in association with
UNESCO." Includes bibliographical
references and index. Subjects:
Biotechnology--Developing countries.
LC Classification: TP248.195.D48 B55
1992 Dewey Class No.:
338.4/76606/091724 20

Biotechnology: economic and wider
impacts. Published/Created: Paris:
Organisation for Economic Co-
operation and Development;
[Washington, D.C.: OECD Publications

and Information Centre, distributor],
1989. Description: 111 p.: ill.; 23 cm.
ISBN: 9264131965 Notes: Published in
French under the Biotechnologie: effets
economiques et autres repercussions.
Includes bibliographical references (p.
103-106). Subjects: Biotechnology--
Economic aspects. Biotechnology--
Social aspects. Biotechnology
industries. LC Classification:
HD9999.B442 B568 1989

Biotechnology: perspectives, policies, and
issues: an international symposium, held
at the University of Florida, Gainesville,
Florida, June 1-4, 1986 / edited by Indra
K. Vasil. Published/Created:
Gainesville: [Published for the]
University of Florida Press [by]
University Presses of Florida, c1987.
Related Authors: Vasil, I. K. University
of Florida. Institute of Food and
Agricultural Sciences. Description: 247
p.: port.; 24 cm. ISBN: 0813008832
Notes: "Sponsored by the Institute of
Food and Agricultural Sciences,
University of Florida; the IC Institute,
The University of Texas at Austin; and
the RGK Foundation." Includes
bibliographies. Subjects:
Biotechnology--Congresses. LC
Classification: TP248.14 .B57 1987
Dewey Class No.: 303.4/83 19

Biotechnology: principles, patents, and
promises, January 23, 1987 / sponsored
by Advanced Legal Education, Hamline
University School of Law.
Published/Created: St. Paul, Minn.
(1536 Hewitt Ave., St. Paul 55104):
ALE, [c1986] Related Authors:
Hamline University. Advanced Legal
Education. Description: 83 leaves; 26
cm. Subjects: Biotechnology industries-
-Law and legislation--United States.
Biotechnology--United States--Patents.

LC Classification: KF1893.B56 B57 1986 Dewey Class No.: 343.73/0786606 347.303786606 19

Biotechnology: professional issues and social concerns / editors, Paul DeForest ... [et al.]. Published/Created: Washington, DC: American Association for the Advancement of Science, Committee on Scientific Freedom and Responsibility, c1988. Related Authors: DeForest, Paul. Description: iv, 111 p.: ill.; 26 cm. Notes: Includes bibliographical references. Subjects: Biotechnology--Moral and ethical aspects. Biotechnology--Social aspects. Series: AAAS publication; 88-23 LC Classification: Q181.A1 A68 no. 88-23 TP248.2 Dewey Class No.: 500 s 174/.966 20

Bisbee, Chester Allan. Bioinformatics: biotechnology enters the information age / author, Chester A. Bisbee. Published/Created: Southborough, MA: International Business Communications, c1997. Description: 1 v. (various pagings): ill.; 29 cm. ISBN: 1579360963 Notes: "October 1997." Subjects: Bioinformatics. Biotechnology industries. Genetic engineering industry. Series: D & MD reports; report #964 LC Classification: R858 .B566 1997

Blakely, Edward James, 1938- Choosing a strategy for local industry development from biotechnology: transfer or incubate? / Edward J. Blakely & Kelvin W. Willoughby. Published/Created: [Berkeley]: Institute of Urban and Regional Development, University of California at Berkeley, [1990] Related Authors: Willoughby, Kelvin W. University of California, Berkeley. Institute of Urban & Regional

Development. Description: i, 39 p.: ill.; 28 cm. Notes: "May 1990." Includes bibliographical references (p. 37-39). Subjects: Biotechnology. Economic policy. Series: Working paper (University of California, Berkeley. Institute of Urban & Regional Development); no. 520. Variant Series: Working paper; 520 LC Classification: MLCM 94/16600 (H)

Blakely, Edward James, 1938- Reformulating the incubator model: applications to commercial biotechnology / Edward J. Blakely & Nancy Nishikawa. Published/Created: Berkeley, Calif.: University of California at Berkeley, Institute of Urban and Regional Development, [1991] Related Authors: Nishikawa, Nancy. University of California, Berkeley. Institute of Urban & Regional Development. Description: 41 p.: ill.; 28 cm. Notes: "This paper was presented in a slightly altered form at the ACSP Conference in Austin, Texas, November 1990." "January 1991." Includes bibliographical references (p. 39-41). Subjects: Biotechnology industries. Economic development. Business incubators. Biotechnology industries-- Government policy. Economic policy. Series: Working paper (University of California, Berkeley. Institute of Urban & Regional Development); no. 528. Variant Series: Working paper; 528 LC Classification: HD9999.B442 B62 1991

California. Legislature. Assembly. Committee on Economic Development and New Technologies. The future of the biotechnology industries in California: Toland Hall Auditorium, University of California-San Francisco Hospital Building, San Francisco, California, June 28, 1984: committee report and record of hearing / Assembly, California Legislature, Assembly Committee on Economic Development and New Tehnologies. Published/Created: [Sacramento, CA]: The Committee: May be purchased from Joint Publications Office, [1984] Description: 18, [37] p.; 28 cm. Notes: "June 28, 1984." Subjects: Biotechnology industries--California. LC Classification: KFC10.4 .E25 1984 Dewey Class No.: 338.4/76606/09794 19

California. Legislature. Assembly. Office of Research. Review of federal and state regulations affecting the California biotechnology industry / prepared by the Assembly Office of Research. Published/Created: Sacramento, Calif.: The Office: Copies from Joint Publications Office, [1985] Related Titles: Biotechnology, a regulatory review. Description: 1 v. (various pagings): ill.; 28 cm. Notes: Cover Biotechnology, a regulatory review. "April 1985." "068-A"--Cover. Includes bibliography. Subjects: Biotechnology industries--Law and legislation-- California. Genetic engineering--Law and legislation--California. Biotechnology industries--Law and legislation--United States. Genetic engineering--Law and legislation-- United States. LC Classification: KFC412.B56 A25 1985

California. Legislature. Senate. Select Committee on Genetics and Public Policy. Truth and consequences of the genetic revolution / Senate Select Committee on Genetics and Public Policy. Published/Created: Sacramento, CA: Senate Publications, 1996. Description: 3 v.; 28 cm. Contents: pt. I. Human genetic research -- pt. II. DNA

on trial -- pt. III. Animal pharm: Old MacDonald had a lab. Notes: Hearings conducted April 8, May 9, and June 24, 1996. "December 1996." "895-S"-- Cover. Includes bibliographical references. Subjects: Medical genetics-- Government policy--California. Genetic engineering--Government policy-- California. DNA fingerprinting-- Government policy--California. Biotechnology--Government policy-- California. LC Classification: RB155 .C35 1996 Dewey Class No.: 616/.042/09794 21

Canada. Ministry of State, Science and Technology. Biotechnology in Canada. Published/Created: [Ottawa, Ont.]: Ministry of State, Science and Technology Canada, 1980. Description: ii, 62 p.; 28 cm. Subjects: Microbial biotechnology--Canada. Industrial microbiology--Canada. Series: MOSST background paper; 11. Variant Series: MOSST background paper; 11 LC Classification: QR53 .C32 1980 Dewey Class No.: 660/.63/0971 19

Canada. Parliament. House of Commons. Standing Committee on Environment and Sustainable Development. Biotechnology regulation in Canada: a matter of public confidence: report of the Standing Committee on Environment and Sustainable Development / Standing Committee on Environment and Sustainable Development. Added TP Réglementation de la biotechnologie au Canada Published/Created: [Ottawa: Queen's Printer for Canada], 1996. Description: x, 56, 58, x p.; 28 cm. Notes: "November 1996." Text in English and French; French text on inverted pages. Subjects: Biotechnology industries--Law and legislation--

Canada. Biotechnology--Research--Law and legislation--Canada. LC Classification: KE1858.B56 A2 1998 Dewey Class No.: 343.71/0786606 21

Canada. Task Force on Biotechnology. Biotechnology: a development plan for Canada: report of the Task Force on Biotechnology to the Minister of State for Science and Technology. Published/Created: [Ottawa]: Minister of Supply and Services Canada, c1981. Description: vii, 51 p.; 24 cm. ISBN: 0662114817 (pbk.) Notes: "Cat. no. ST 31-9/1981E"--T.p. verso. Includes bibliographical references. Subjects: Biotechnology--Canada--Planning. LC Classification: TP248.19.C2 C36 1981

Changing nature's course: the ethical challenge of biotechnology / edited by Gerhold K. Becker in association with James P. Buchanan. Published/Created: Hong Kong: Hong Kong University Press, c1996. Related Authors: Becker, Gerhold K., 1943- Buchanan, James Porter, 1948- Description: x, 208 p.: ill.; 25 cm. ISBN: 9622094031 Notes: Articles out of the symposium "Biotechnology and Ethics: Scientific Liberty and Moral Responsibility" held in November of 1993; organized and sponsored by the Centre for Applied Ethics at Hong Kong Baptist University in cooperation with the Hong Kong University of Science and Technology and the Goethe Institut. Includes bibliographical references and index. Subjects: Biotechnology--Moral and ethical aspects--Congresses. LC Classification: TP248.14 .C47 1996 Dewey Class No.: 174/.966/06 21

Conference to Promote Japan/U.S. Joint Projects and Cooperation in Biotechnology (3rd: 1991: Honolulu, Hawaii) Bioproducts and bioprocesses 2 / Third Conference to Promote Japan/U.S. Joint Projects and Cooperation in Biotechnology, Honolulu, Hawaii, January 6-10, 1991; T. Yoshida, R.D. Tanner, eds. Published/Created: Berlin; New York: Springer-Verlag, c1993. Related Authors: Yoshida, T. (Toshiomi), 1939- Tanner, R. D. (Robert D.) Related Titles: Bioproducts and bioprocesses two. Description: xiii, 308 p.: ill.; 25 cm. ISBN: 3540565086 (Berlin: acid-free paper) 0387565086 (New York: acid-free paper) Notes: Includes bibliographical references. Subjects: Biochemical engineering--Congresses. Biotechnology--Congresses. LC Classification: TP248.3 .C65 1991 Dewey Class No.: 660/.6 20

Coombs, Joseph E., 1966- International strategy and market performance in new biotechnology firms / Joseph E. Coombs. Published/Created: New York: Garland Pub., 1999. Description: xiii, 94 p.; 23 cm. ISBN: 0815335016 (alk. paper) Notes: Originally presented as the author's thesis (doctoral--Temple University). Includes bibliographical references (p. 71-90) and index. Subjects: Biotechnology industries. New business enterprises. International business enterprises. International trade. Export marketing. Market surveys. Series: Transnational business and corporate culture LC Classification: HD9999.B442 C663 1999 Dewey Class No.: 338.4/76606 21

Cooper, Iver P. Biotechnology and the law / by Iver P. Cooper. Published/Created: New York, N.Y.: C. Boardman Co., 1982- Description: 2 v. (loose-leaf); 26 cm. ISBN: 0876323115 Notes: Includes bibliographical references and index.

Subjects: Microorganisms--United States--Patents. Biotechnology--United States--Patents. Series: Intellectual property library. Variant Series: The Clark Boardman intellectual property library LC Classification: KF3133.B56 C66 1982 Dewey Class No.: 346.04/86 342.6486 19

Coors, Marilyn E., 1947- The matrix: charting an ethics of inheritable genetic modification / Marilyn E. Coors. Published/Created: Lanham: Rowman & Littlefield, 2002. Projected Pub. Date: 0210 Description: p. cm. ISBN: 0742514005 (cloth: alk. paper) 0742514013 (pbk.: alk. paper) Notes: Includes bibliographical references and index. Subjects: Genetic engineering--Moral and ethical aspects. Biotechnology--Moral and ethical aspects. LC Classification: QH438.7 .C665 2002 Dewey Class No.: 174/.966 21

Crespi, R. S. Patents: a basic guide to patenting in biotechnology / R.S. Crespi. Published/Created: Cambridge [Cambridgeshire]; New York: Cambridge University Press, 1988. Description: vii, 191 p.: ill.; 24 cm. ISBN: 052132954X Notes: Includes index. Bibliography: p. 188. Subjects: Biotechnology--Patents. Biotechnology--United States--Patents. Series: Cambridge studies in biotechnology; 6 LC Classification: K1519.B54 C74 1988 Dewey Class No.: 346.04/86 342.6486 19

Critser, James R. Biotechnical engineering: equipment and processes: retrieval index and abstracts of U.S. patents / prepared by James R. Critser, Jr. Published/Created: Ashland, Mass.: Lexington Data, c1990. Description: iii,

651 p.: ill; 28 cm. ISBN: 0881780723 Subjects: Biotechnology--Patents--United States--Indexes. Biochemical engineering--Patents--United States--Indexes. Series: Report (Lexington Data, Inc.); 89-14 Variant Series: Report /Lexington Data; #14-89 LC Classification: TP248.24 .C75 1990 Dewey Class No.: 660/.6/027273 20

Crueger, Wulf. Biotechnology: a textbook of industrial microbiology / Wulf Crueger, Anneliese Crueger; editor of the English edition, Thomas D. Brock. English Edition Information: 2nd ed. Published/Created: Sunderland, MA: Sinauer Associates, c1990. Related Authors: Crueger, Anneliese. Brock, Thomas D. Description: x, 357 p.: ill.; 25 cm. ISBN: 0878931317: Notes: Translation of: Lehrbuch der angewandten Mikrobiologie. Includes bibliographical references. Subjects: Industrial microbiology. Microbial biotechnology. LC Classification: QR53 .C7813 1990 Dewey Class No.: 660/.62 20

Dadzie, Stephen. Biotechnology in sub-Saharan Africa: policy and institutional options / Stephen Dadzie. Published/Created: Nairobi, Kenya: Acts Press, 2001. Description: vii, 38 p.; 21 cm. ISBN: 9966411143 Notes: Includes bibliographical references (p. 35-38). Subjects: Biotechnology industries--Government p[olicy--Africa, Sub-Saharan. Series: Science and technology policy discussion paper; no. 1 LC Classification: HD9999.B443 A3573 2001 Dewey Class No.: 338.4/76606/0967 21

Daly, Peter, 1953- The biotechnology business: a strategic analysis / Peter Daly. Published/Created: Totowa, NJ:

Rowman & Allanheld, 1985. Description: 155 p.; 23 cm. ISBN: 0847674606: Notes: Includes index. Bibliography: p. 138-143. Subjects: Biotechnology industries. Market surveys. Biomedical Engineering. Industry. Technology, Medical. LC Classification: HD9999.B442 D35 1985 Dewey Class No.: 338.4/76208 19

Daly, Peter, 1953- The biotechnology business: a strategic analysis / Peter Daly. Published/Created: London: F. Pinter; Totowa, N.J.: Rowman & Allanheld, 1985. Description: 155 p.; 23 cm. ISBN: 0847674606 0847674606 (Rowman & Allanheld): Notes: Includes index. Bibliography: p. 138-143. Subjects: Biotechnology industries. Market surveys. LC Classification: HD9999.B442 D35 1985b Dewey Class No.: 338.4/76208 19

Dodgson, Mark, 1957- The management of technological learning: lessons from a biotechnology company / Mark Dodgson. Published/Created: Berlin; New York: W. de Gruyter, 1991. Description: ix, 150 p.: ill.; 24 cm. ISBN: 0899257682 (U.S.: acid-free paper) 3110127067 (acid-free paper) Notes: Includes bibliographical references (p. [137]-141) and index. Subjects: Celltech Ltd.--Management. Biotechnology industries--Great Britain--Management--Case studies. Technological innovations--Employee participation--Great Britain--Case studies. Series: De Gruyter studies in organization; 29 LC Classification: HD9999.B444 C453 1991 Dewey Class No.: 660/.6/068 20

Dronamraju, Krishna R. Biological and social issues in biotechnology sharing / Krishna R. Dronamraju.

Published/Created: Aldershot; Brookfield: Ashgate Pub., c1998. Description: x, 165 p.: ill.; 23 cm. ISBN: 1840148977 Notes: Includes bibliographical references (p. 157-162) and index. Subjects: Biotechnology--Patents. Biotechnology--Technology transfer. LC Classification: TP248.175 .D76 1998 Dewey Class No.: 660.6 21

Duarte, Joe. Successful biotech investing: every investor's complete guide / Joe Duarte. Published/Created: Roseville, Calif.: Prima Money, c2001. Description: v, 409 p.: ill.; 24 cm. ISBN: 076153301X Notes: Includes bibliographical references (p. 389-390) and index. Subjects: Biotechnology industries--Finance. Investments. Stocks. LC Classification: HD9999.B442 D8 2001 Dewey Class No.: 332.63/22 21

Ducor, Philippe Georges. Patenting the recombinant products of biotechnology / by Philippe Georges Ducor. Published/Created: London; Boston: Kluwer Law International, 1998. Description: xvii, 179 p.: ill.; 25 cm. ISBN: 9041106987 Notes: Includes bibliographical references (p. 167-172) and index. Subjects: Patent laws and legislation--United States. Biotechnology industries--Law and legislation--United States. Recombinant DNA--Patents. LC Classification: KF3133.B56 D83 1998

Dullea, Mark. Marine biopolymers / Mark Dullea, project analyst. Published/Created: Norwalk, CT: Business Communications Co., [1994] Related Authors: Business Communications Co. Description: xxi, 159 leaves; 28 cm. ISBN: 1569650268 Notes: "August 1994"--Verso t.p.

Subjects: Biotechnology industries. Biopolymers--Industrial applications. Market surveys. Series: Business opportunity report; C-184A LC Classification: HD9999.B442 D85 1994

Dullea, Mark. New developments in marine biotechnology / Mark Dullea, project analyst. Published/Created: Norwalk, CT: Business Communications Co., [1994] Related Authors: Business Communications Co. Description: xxii, 218 leaves: ill.; 28 cm. ISBN: 1569650276 Notes: "September 1994"-- Verso t.p. Includes bibliographical references (p. 205-218). Subjects: Aquaculture industry. Marine biotechnology. Biotechnology industries. Market surveys. Series: Business opportunity report; C-184B LC Classification: HD9450.5 .D85 1994

Elkington, John. Bio-Japan: the emerging Japanese challenge in biotechnology / by John Elkington. Published/Created: London: Oyez Scientific & Technical Services, c1985. Description: ix, 109 p.: ill.; 30 cm. ISBN: 0907822525 (pbk.): Subjects: Biotechnology industries-- Japan. LC Classification: HD9999.B443 J34 1985 Dewey Class No.: 338.4/76606/0952 19

Genome valley: the economic potential and strategic importance of biotechnology in the UK: report / [DTI]. Published/Created: [London]: DTI, 1999. Related Authors: Great Britain. Dept. of Trade and Industry. Description: vi, 64 p.: map; 30 cm. Notes: Cover title. Includes bibliographical references. Subjects: Biotechnology industries--Great Britain. Genetic engineering industry--Great Britain. LC Classification:

HD9999.B443 G74 1999

Graham, Elaine L. Representations of the post/human: monsters, aliens, and others in popular culture / Elaine L. Graham. Published/Created: New Brunswick, NJ: Rutgers University Press, 2002. Projected Pub. Date: 0204 Description: p. cm. ISBN: 081353058X (alk. paper) 0813530598 (pbk.: alk. paper) Notes: Includes bibliographical references and index. Subjects: Technology--Social aspects. Biotechnology--Social aspects. Cybernetics--Social aspects. Popular culture. Postmodernism. LC Classification: HM846 .G73 2002 Dewey Class No.: 306.4/6 21

Griffing, John M. Silicon Valley II: a review of state biotechnology development incentives / prepared by Senate Office of Research. Published/Created: [Sacramento, Calif.]: Senate Office of Research: Copies from Joint Publications Office, [1985] Related Authors: California. Legislature. Senate. Office of Research. Related Titles: Silicon Valley 2. Silicon Valley two. Description: 2 leaves, iii, 58 p.; 28 cm. Notes: Cover title. "August 1985." "134-S." Bibliography: p. [53]-56. Subjects: Biotechnology industries-- Government policy--California. LC Classification: HD9999.B443 U64 1985 Dewey Class No.: 338.4/76606/09794 19

Guenther, Kim. Biotechnology, legislation and regulation: January 1989 - January 1994 / Kim Guenther and Raymond Dobert. Published/Created: Beltsville, Md.: National Agricultural Library, [1994] Related Authors: Dobert, Raymond. National Agricultural Library (U.S.) Description: 47 p.; 28 cm. Notes: "297 citations in English from

AGRICOLA." "Updates QB 92-53."
Shipping list no.: 94-0214-P. "May
1994." Includes indexes. Subjects:
Agricultural biotechnology--Law and
legislation--United States--
Bibliography. Biotechnology--
Research--Law and legislation--United
States--Bibliography. Series: Quick
bibliography series; 94-33. Variant
Series: Quick bibliography series, 1052-
5378; QB 94-33 LC Classification:
KF1893.B56 A124 1994

Hacking, Andrew J. Economic aspects of
biotechnology / Andrew J. Hacking.
Published/Created: Cambridge
[Cambridgeshire]; New York:
Cambridge University Press, 1986.
Description: x, 306 p.: ill.; 24 cm.
ISBN: 0521258936 Notes: Includes
index. Bibliography: p. 289-296.
Subjects: Biotechnology industries.
Series: Cambridge studies in
biotechnology; 3 LC Classification:
HD9999.B442 H33 1986 Dewey Class
No.: 338.4/76208 19

Howells, John, 1961- Managers and
innovation: strategies for a
biotechnology / John Howells.
Published/Created: London; New York:
Routledge, 1994. Description: 241 p.;
23 cm. ISBN: 041508590X Notes:
Includes bibliographical references (p.
[216]-224) and index. Subjects:
Biotechnology industries.
Biotechnology--Industrial applications--
Research Management. Artificial foods-
-Research--Management. Single cell
proteins--Research--Management. LC
Classification: HD9999.B442 H69 1994
Dewey Class No.: 338.4/56606 20

India. Dept. of Science and Technology.
Performance budget.
Published/Created: New Delhi: Ministry

of Science and Technology, Dept. of
Science and Technology & Dept. of
Scientific & Industrial Research,
Related Authors: India. Dept. of
Scientific & Industrial Research. India.
Dept. of Biotechnology. Description: v.;
30 cm. Current Frequency: Annual
Notes: Description based on: 1986-87.
Vols. for 1987-88- issued with Dept. of
Biotechnology. SERBIB/SERLOC
merged record Subjects: India. Dept. of
Science and Technology--
Appropriations and expenditures--
Periodicals. India. Dept. of Scientific &
Industrial Research Appropriations and
expenditures--Periodicals. India. Dept.
of Biotechnology--Appropriations and
expenditures--Periodicals. LC
Classification: Q127.I4 I47c

India. Ministry of Science and Technology.
Performance budget [microform] /
Ministry of Science and Technology.
Published/Created: New Delhi: The
Ministry, Description: v.; 29 cm.
Current Frequency: Annual Continues:
India. Dept. of Science and Technology.
Performance budget Notes: Master
microform held by: DLC. Vols. for
include performance budgets for Dept.
of Science and Technology, Dept. of
Scientific & Industrial Research, and,
Dept. of Biotechnology. Description
based on: 1992-93. Microfiche.
1992/93- New Delhi: Library of
Congress Office; Washington, D.C.:
Library of Congress Photoduplication
Service, 1995- microfiches.
SERBIB/SERLOC merged record
Subjects: India. Dept. of Science and
Technology--Appropriations and
expenditures--Periodicals. India. Dept.
of Scientific & Industrial Research
Appropriations and expenditures--
Periodicals. India. Dept. of
Biotechnology--Appropriations and

expenditures--Periodicals. LC Classification: Microfiche (o) 95/60160 (Q) Overseas Acq. No.: I-E-E-sf 95072289

Kameri-Mbote, Patricia. Public/private partnerships for biotechnology in Africa: the future agenda / Patricia Kameri-Mbote, David Wafula, and Norman Clark. Published/Created: Nairobi, Kenya: Acts Press, 2001. Related Authors: Wafula, David. Clark, Norman. Description: v, 43 p.; 21 cm. ISBN: 9966411151 Notes: Includes bibliographical references (p. 40-43). Subjects: Biotechnology--Africa. Biotechnology industries--Africa. Series: ACTS/BIO-EARN occasional paper; 1 LC Classification: TP248.195.A35 K35 2001

Karossi, Tenri A. Pengembangan dan pemanfaatan bioteknologi untuk mengurangi ketergantungan kepada produk impor dan meningkatkan kandungan lokal / oleh A.T. Karossi. Published/Created: [Jakarta]: Departemen Pertahanan Keamanan RI, Lembaga Ketahanan Nasional, 1998. Description: v, 92, [1] leaves; 29 cm. Notes: Includes bibliographical references (leaf [93]). Subjects: Biotechnology industries; policy; Indonesia Variant Series: Kertas karya perorangan (TASKAP) kursus singkat LC Classification: MLCME 2000/02307 (H)

Lanzavecchia, Giuseppe. The impact of biotechnology on working conditions / [by Guiseppe [sic] Lanzavecchia, Danielle Mazzonis]. Published/Created: Loughlinstown House, Shankill, Co. Dublin, Ireland: European Foundation for the Improvement of Living and Working Conditions, c1987. Related

Authors: Gattegno Mazzonis, Danielle. Description: x, 105 p.; 30 cm. ISBN: 9282567672 Notes: Includes bibliographical references (p. 71-83). Subjects: Work environment. Industrial hygiene. Biotechnology industries. Biotechnology. Series: Research report (European Foundation for the Improvement of Living and Working Conditions) Variant Series: Research report LC Classification: HD6955 .L32 1987 Dewey Class No.: 331.25 20

Legal aspects of biotechnology: proceedings of international symposium, October 1986 / chaired by Shoshana Berman; edited by Paulina Ben-Ami. Published/Created: Jerusalem, Israel: Ministry of Science and Technology, [1989?] Related Authors: Berman, Shoshana. Ben-Ami, Paulina. Description: 130 p.; 30 cm. Notes: Cover title. "NCRD 89-12." Includes bibliographical references. Subjects: Biotechnology industries--Law and legislation--Congresses. LC Classification: K3925.B56 A55 1986 Dewey Class No.: 341.7/6754 20

McCarthy, Wil. Bloom / Wil McCarthy. Edition Information: 1st ed. Published/Created: New York: Ballantine Pub. Group, 1998. Description: 310 p.; 24 cm. ISBN: 0345408578 (alk. paper) Notes: "A Del Rey book"--T.p. verso. Subjects: Biotechnology--Fiction. Genre/Form: Science fiction. LC Classification: PS3563.C337338 B58 1998 Dewey Class No.: 813/.54 21

National biotech register. Published/Created: Wilmington, MA: Barry, c1992- Description: v.; 23 cm. Vol. 1- Current Frequency: Annual ISSN: 1074-9942 Notes: Title from cover.

SERBIB/SERLOC merged record Subjects: Biotechnology industries--United States--Directories. Biotechnology industries--United States--Registers. Biotechnology--United States--directories. Industry--directories. LC Classification: HD9999.B443 U658 Dewey Class No.: 338.7/6606/029473 20

NATO Advanced Study Institute on Recent Developments in Biotechnology (1985: Troia, Portugal) Perspectives in biotechnology / edited by J.M. Cardoso Duarte ... [at al.]. Published/Created: New York: Plenum Press, c1987. Related Authors: Duarte, J. M. Cardoso (José M. Cardoso) North Atlantic Treaty Organization. Scientific Affairs Division. Description: x, 207 p.: ill.; 26 cm. ISBN: 0306425696 Notes: "Published in cooperation with NATO Scientific Affairs Division." Includes bibliographies and index. Subjects: Biotechnology--Congresses. Series: NATO ASI series. Series A, Life sciences; v. 128 LC Classification: TP248.14 .N38 1985 Dewey Class No.: 660/.6 19

Nature, risk, and responsibility: discourses of biotechnology / edited by Patrick O'Mahony; consultant editor, Jo Campling. Published/Created: New York: Routledge, 1999. Related Authors: O'Mahony, Patrick, 1957- Description: x, 232 p.: ill.; 23 cm. ISBN: 0415922909 (hbk.) 0415922917 (pbk.) Notes: Includes bibliographical references and index. Subjects: Biotechnology--Social aspects. Biotechnology--Moral and ethical aspects. LC Classification: TP248.23 .N38 1999 Dewey Class No.: 174/.96606 21

New biotechnology comes to market / Philip Rotheim, project analyst. Published/Created: Stamford, Conn., U.S.A.: Business Communications Co., c1985. Related Authors: Rotheim, Philip. Business Communications Co. Description: viii, 205 leaves: ill.; 29 cm. ISBN: 0893364363 (soft) Notes: "July 1985"--T.p. verso. Subjects: Biotechnology industries. Market surveys. Series: Business opportunity report; C-032R LC Classification: HD9999.B442 N48 1985 Dewey Class No.: 381/.456606 19

New developments in biotechnology. Published/Created: Washington, DC: Congress of the U.S., Office of Technology Assessment: For sale by the Supt. of Docs., U.S. G.P.O., 1987-<1989 Related Authors: United States. Congress. Office of Technology Assessment. Description: v. <1-5: ill., forms; 26 cm. Incomplete Contents: v. 1. Ownership of human tissues and cells -- v. 2. Public perceptions of biotechnology -- v. 3. Field-testing engineered organisms -- v. 4. U.S. investment in biotechnology -- v. 5. Patenting life. Notes: Shipping list no.: 87-185-P (v. 1) Shipping list no.: 87-319-P (v. 2) "OTA-BA-337 March 1987"--Vol. 1, p. [4] of cover. "OTA-BP-BA-45 May 1987"--Vol. 2, p. [4] of cover. "OTA-BA-350 May 1988"--Vol. 3, p. [4] of cover. "OTA-BA-360 July 1988"--Vol. 4, p. [4] of cover. "OTA-BA-370 April 1989"-- Vol. 5, p. [4] of cover. S/N 052-003-01060-7 (v.1) S/N 052-003-01068-2 (v.2) Item 1070-M Includes bibliographies and indexes. Subjects: Biotechnology--United States--Public opinion. Biotechnology. Series: Background paper (United States. Congress. Office of Technology Assessment) Variant Series: v. 2:

Background paper LC Classification: TP248.185 .N482 1987 Dewey Class No.: 660/.6 19

New developments in biotechnology: patenting life / Office of Technology Assessment, Congress of the United States. Published/Created: New York: Dekker, c1990. Related Authors: United States. Congress. Office of Technology Assessment. Related Titles: Patenting life. Description: vii, 195 p.: ill.; 29 cm. ISBN: 0824780701 (alk. paper) Notes: Reprint. Originally published: New developments in biotechnology, v. 5: Washington, D.C.: Congress of the U.S., Office of Technology Assessment, 1987. Subjects: Biotechnology--United States--Patents. LC Classification: KF3133.B56 N49 1990 Dewey Class No.: 343.73/0766606 347.303766606 20

New England Biotechnology Law Conference '94 (1994) Biotech '94 / New England Biotechnology Law Conference '94; Peter Wirth, cochair ... [et al.]. Published/Created: Boston, MA (10 Winter Pl., Boston 02108-4751): MCLE, c1994. Related Authors: Wirth, Peter, 1950- Massachusetts Continuing Legal Education, Inc. (1982-) Description: xiv, 418 p.: ill.; 22 cm. Notes: "No. 95-13.01." Subjects: Biotechnology industries--Law and legislation--United States--Congresses. Biotechnology industries--Law and legislation--Europe Congresses. LC Classification: KF1893.B56 N48 1994

New England Biotechnology Law Conference '96 (1994) Biotech '96: New England biotechnology law conference '96 / jointly sponsored with the American Law Institute-American Bar Association (ALI-ABA) Committee on Continuing Professional Education;

Brian C. Cunningham, cochair. Published/Created: Boston, MA (10 Winter Pl., Boston 02108-4751): Massachusetts Continuing Legal Education, [c1996] Related Authors: Cunningham, Brian (Brian C.) American Law Institute-American Bar Association Committee on Continuing Professional Education. Massachusetts Continuing Legal Education, Inc. (1982-) Description: xiv, 296 p.: ill.; 22 cm. Notes: "No. 97-13.01." Includes bibliographical references (p. 285). Subjects: Biotechnology industries-- Law and legislation--United States-- Congresses. LC Classification: KF1893.B56 N48 1996 Dewey Class No.: 343.73/0786606 21

New Jersey. Biotechnology Development Task Force. Meeting. (1993: Trenton, N.J.) Meeting of Biotechnology Development Task Force: regulatory environment affecting the biotechnology industry and the promotion of the industry through transfer of technology. Published/Created: Trenton, N.J.: Office of Legislative Services, Public Information Office, Hearing Unit, [1993] Related Titles: Regulatory environment affecting the biotechnology industry and the promotion of the industry through the transfer of technology. Description: 62, 75 p.: ill.; 28 cm. Notes: Cover title. "Issued 11/08/93"--Prelim. p. [i]. Meeting held on November 16, 1993, at Trenton, N.J. Subjects: Biotechnology industries--Government policy--New Jersey Congresses. Biotechnology industries--New Jersey--Finance-- Congresses. LC Classification: HD9999.B443 U659 1993

New Jersey. Biotechnology Development Task Force. Meeting. (1993: Trenton, N.J.) Meeting of Biotechnology Development Task Force: ways in which the state and private sector can encourage the investment in and the financing of the biotechnology industry in New Jersey. Published/Created: Trenton, N.J.: Office of Legislative Services, Public Information Office, Hearing Unit, [1993] Description: 96, 35 p.; 28 cm. Notes: Cover title. Meeting held on October 8, 1993, Trenton, N.J. Subjects: Biotechnology industries--Government policy--New Jersey Congresses. Biotechnology industries--New Jersey--Finance--Congresses. LC Classification: HD9999.B443 U659 1993a

O'Brien, James W. Science policy, biotechnology, and American state government: recommendations for state action / by James W. O'Brien. Published/Created: Ames, Iowa, USA: Technology and Social Change Program, Iowa State University, c1989. Description: 76 p.; 28 cm. ISBN: 0945271174 Notes: Includes bibliographical references (p. 69-76). Subjects: Agricultural biotechnology--Government policy--Iowa. Science and state--Iowa. Series: Studies in technology and social change series, 0896-1905; no. 12 LC Classification: S494.5.B563 O27 1989 Dewey Class No.: 338.1/8777 20

Organisation for Economic Co-operation and Development. Safety considerations for biotechnology, 1992. Published/Created: Paris, France: Organization for Economic Co-operation and Development; Washington, D.C.: OECD Publications and Information Centre [distributor], c1992. Related Authors: Organisation for Economic Co-operation and Development. Group of National Experts on Safety and Biotechnology. Description: 50 p.: ill.; 23 cm. ISBN: 926413641X: Notes: Prepared by the Group of National Experts on Safety and Biotechnology.Cf. Preface. Includes bibliographical references (p. 49-50). Subjects: Biotechnology--Safety measures. LC Classification: TP248.2 .O74 1992 Dewey Class No.: 660/.65/0289 20

Orsenigo, Luigi, 1954- The emergence of biotechnology: institutions and markets in industrial innovation / Luigi Orsenigo. Published/Created: London: Pinter, 1989. Description: xi, 230 p.: ill.; 24 cm. ISBN: 0861877020 Notes: Includes bibliographical references (p. [208]-225) and index. Subjects: Biotechnology industries--History. LC Classification: HD9999.B442 O77 1989b

Orsenigo, Luigi, 1954- The emergence of biotechnology: institutions and markets in industrial innovation / Luigi Orsenigo. Published/Created: New York: St. Martin's Press, 1989. Description: xi, 230 p.: ill.; 24 cm. ISBN: 0312031971 Notes: Includes index. Bibliography: p. [208]-225. Subjects: Biotechnology industries--History. LC Classification: HD9999.B442 O77 1989 Dewey Class No.: 660/.6 20

Pappas, Michael G., Ph. D. The biobusiness handbook: how to organize and operate a biotechnology business, including the most promising applications for the 1990s / by Michael G. Pappas. Published/Created: Totowa, NJ: Humana Press, c1994. Description: 461

p.: ill.; 29 cm. ISBN: 0896032183
Notes: Loose-leaf in binder. Includes
bibliographical references and index.
Subjects: Biotechnology industries--
Management. New business enterprises-
-Management. LC Classification:
HD9999.B442 P36 1993 Dewey Class
No.: 660.6/068 20

Plants, power, and profit: social, economic,
and ethical consequences of the new
biotechnologies / [edited by] Lawrence
Busch ... [et al.]. Published/Created:
Cambridge, Mass., USA: B. Blackwell,
1991. Related Authors: Busch,
Lawrence. Description: xii, 275 p.: ill.;
24 cm. ISBN: 1557860882: Notes:
Includes bibliographical references (p.
[239]-264) and index. Subjects:
Biotechnology industries.
Biotechnology industries--Social
aspects. Biotechnology industries--
Moral and ethical aspects. LC
Classification: HD9999.B442 B87 1991
Dewey Class No.: 338.4/76606 20

Preclinical safety of biotechnology products
intended for human use: proceedings of
the satellite symposium to the IV
International Congress of Toxicology,
Keio Plaza Hotel, Tokyo, Japan, July
26, 1986 / editor, Charles E. Graham.
Published/Created: New York: A.R.
Liss, c1987. Related Authors: Graham,
Charles E. International Congress on
Toxicology (4th: 1986: Tokyo, Japan)
Description: xv, 213 p.: ill.; 24 cm.
ISBN: 0845150855 Notes: Includes
bibliographies and index. Subjects:
Biotechnology--Safety measures--
Congresses. Biotechnology--Safety
regulations--Congresses. Toxicology--
Congresses. Biotechnology--congresses.
Series: Progress in clinical and
biological research; v. 235 LC
Classification: TP248.14 .P74 1987

Dewey Class No.: 363.1/94 19

Rainis, Kenneth G. Biotechnology projects
for young scientists / Kenneth G. Rainis
and George Nassis. Published/Created:
New York: Franklin Watts, c1998.
Related Authors: Nassis, George.
Description: 160 p.: ill.; 24 cm. ISBN:
0531114198 Summary: Gives
instructions for and explains the
principles behind a variety of
biotechnology experiments from the
simple to the more difficult. Notes:
Includes bibliographical references (p.
154-155) and index. Subjects:
Biotechnology projects. Science
projects. Experiments. LC
Classification: TP248.22 R35 1998
Dewey Class No.: 660/.6/078 21

Reichel, Brian J. Research through
biotechnology: institutional impacts and
societal concerns: a guide to the
literature / by Brian J. Reichel, William
F. Woodman, Mack C. Shelley.
Published/Created: Ames, Iowa:
Technology and Social Change
Program, Iowa State University, c1987.
Related Authors: Woodman, William F.
Shelley, Mack C., 1950- Description:
viii, 139 p.; 28 cm. ISBN: 094527100X
Subjects: Biotechnology--Research--
Bibliography. Series: Bibliographies in
technology and social change series,
0896-1689; no. 1 LC Classification:
Z7914.B33 R45 1987 TP248.2 Dewey
Class No.: 016.66/06/072073 20

Rigaux, Fabrice, 1968- Industrial
biotechnology in the Atlantic provinces:
from emergence to development? /
Fabrice Rigaux. Published/Created:
[Moncton, N.B.]: Canadian Institute for
Research on Regional Development =
Institut canadien de recherche sur le
developpement regional, [1997?].

Related Authors: Université de
Moncton. Canadian Institute for
Research on Regional Development.
Description: 139 p.: ill.; 23 cm. ISBN:
0886590507 Notes: Issued also in
French under Les biotechnologies
industrielles dans les provinces
maritimes. Includes bibliographical
references: p. 123-132. Subjects:
Biotechnology industries--Atlantic
Provinces. Biotechnology--Economic
aspects--Atlantic Provinces.
Biotechnology industries--Canada. Bio-
industries--Provinces de l'Atlantique.
Biotechnologie--Aspect économique--
Provinces de l'Atlantique. Bio-
industries--Canada. Atlantic Provinces--
Economic policy. Provinces de
l'Atlantique--Politique économique.
Series: Maritime series. Monographs
LC Classification: ACQUISITION IN
PROCESS (COPIED) (lccopycat)

Safety in industrial microbiology and
biotechnology / edited by C.H. Collins
and A.J. Beale. Published/Created:
Oxford; Boston: Butterworth
Heinemann, 1992. Related Authors:
Collins, C. H. (Christopher Herbert)
Beale, A. J. Description: vi, 257 p.: ill.;
25 cm. ISBN: 0750611057: Notes:
Includes bibliographical references and
index. Subjects: Biotechnology--Safety
measures. Industrial microbiology--
Safety measures. Genetic engineering--
Safety measures. LC Classification:
TP248.2 .S24 1992 Dewey Class No.:
660/.6/0289 20

Science, truth and justice / edited by Joost
Blom, Hélène Dumont; [conference
organized by] CIAJ, Canadian Institute
for the Administration of Justice =
Science, vérité et justice / sous la
direction de Joost Blom, Hélène
Dumont; [conférence organisée par]

ICAJ, Institut canadien d'administration
de la justice. Parallel Science, vérité et
justice Published/Created: Montréal:
Éditions Thémis, 2001. Related
Authors: Blom, Joost. Dumont, Hélène,
1947- Canadian Institute for the
Administration of Justice. Description:
xii, 311 p.; 25 cm. ISBN: 2894001517
Notes: Proceedings of a conference held
in Victoria, B.C., Oct. 11-14, 2000.
"2000". Text in English and French.
Includes bibliographical references.
Subjects: Science and law--Congresses.
Technology and law--Congresses.
Internet--Law and legislation--
Congresses. Danger (Law)--Congresses.
Biotechnology--Law and legislation--
Congresses. Evidence (Law)--
Congresses.

The changing bioreactor business / Cort
Wrotnowski, project analyst.
Published/Created: Norwalk, CT:
Business Communications Co., [1990]
Related Authors: Wrotnowski, Cort.
Business Communications Co.
Description: xviii, 173 leaves: ill.; 28
cm. ISBN: 089336780X Notes: "July
1990"--T.p. verso. Subjects: Biological
laboratory equipment industry.
Biological apparatus and supplies.
Biotechnology. Market surveys. Series:
Business opportunity report; C-086 LC
Classification: HD9706.65.B552 C48
1990

The Impact of biotechnology on living and
working conditions: a selected
bibliography / edited by V. Di Martino,
E. Yoxen, and R. Wakeford.
Published/Created: Shankill, Co.
Dublin, Ireland: European Foundation
for the Improvement of Living and
Working Conditions; Washington, DC:
European Community Information
Service [distributor], c1987. Related

Authors: Di Martino, Vittorio. Yoxen, Edward. Wakeford, R. (Richard) European Foundation for the Improvement of Living and Working Conditions. Description: vii, 107 p.; 30 cm. ISBN: 9282570649 Notes: On verso of t.p.: Luxembourg: Office for Official Publications of the European Communities (Catalogue number: SY-48-47-404-EN-C). Includes indexes. Subjects: Biotechnology--Social aspects--Bibliography. LC Classification: Z7914.S663 I46 1987

The Legal aspects of commercializing medical high-tech/biotech products. Published/Created: Boston, Mass. (20 West St., Boston 02111): Massachusetts Continuing Legal Education, c1989. Related Authors: Massachusetts Continuing Legal Education, Inc. (1982-) Description: xxii, 386 p.; 24 cm. Notes: Page 386 blank. "(89-120)." Subjects: Biotechnology industries--Law and legislation--United States. Biotechnology--Research--Law and legislation--United States. Biotechnology--United States--Patents. Biotechnology industries--Law and legislation Massachusetts. Biotechnology--Research--Law and legislation Massachusetts. LC Classification: KF1893.B56 L44 1989 Dewey Class No.: 343.73/0786606 347.303786606 20

The management and economic potential of biotechnology / [guest editors] Edward J. Blakely, Kelvin W. Willoughby. Published/Created: Geneva: Inderscience Enterprises, c1993. Related Authors: Blakely, Edward James, 1938- Willoughby, Kelvin W. Related Titles: [Journal of technology management. Description: 191 p.: ill. Cancelled ISBN: 0907776122 Notes:

"A special publication of the Internaitonal journal of technology management"--Cover. Includes bibliographical references. Subjects: Biotechnology--organization & administration. Biotechnology--economics. Biotechnology. Biotechnology industries--Economic aspects. Series: Biotechnology review (Geneva, Switzerland); no. 1. Variant Series: Biotechnology review; no. 1

The performance of distributors and prime vendors in the North American life sciences market: final report. Published/Created: Marblehead, MA: Life Sciences Marketing Group, [c1997] Related Authors: Thompson, J. A. (Judy A.) Life Sciences Marketing Group. Description: 96 leaves: ill.; 29 cm. Notes: "An activity of J.A. Thompson." Subjects: Biotechnology industries--United States. Pharmaceutical industry--United States. Market surveys--United States. LC Classification: HD9999.B443 U6617 1997

Toyota Conference (4th: 1990: Aichi-ken, Japan) Automation in biotechnology: a collection of contributions presented at the Fourth Toyota Conference, Aichi, Japan, 21-24 October 1990 / edited by Isao Karube. Published/Created: Amsterdam; New York: Elsevier; New York, NY, U.S.A.: Distributors for the U.S. and Canada Elsevier Science Pub. Co., c1991. Related Authors: Karube, Isao, 1942- Description: xvii, 386 p.: ill.; 25 cm. ISBN: 0444887679 (acid-free paper) Subjects: Biotechnology--Automation--Congresses. LC Classification: TP248.25.A96 T68 1990 Dewey Class No.: 660/.6 20

Trends in biotechnology and chemical patent practice, 1989 / chairman, Nels

T. Lippert. Published/Created: New York, N.Y. (810 Seventh Ave., New York 10019): Practising Law Institute, c1989. Related Authors: Lippert, Nels T. Description: 232 p.; 22 cm. Notes: "Prepared for distribution at the Trends in biotechnology and chemical patent practice program, December 11-12, 1989, New York City"--P. 5. "G4-3843." Subjects: Biotechnology industries--Law and legislation--United States. Biotechnology--Research--Law and legislation--United States. Biotechnology--United States--Patents. Chemicals--United States--Patents. Series: Patents, copyrights, trademarks, and literary property course handbook series; no. 286 LC Classification: KF1893.B56 T74 1989 Dewey Class No.: 343.73/07862082 347.3037862082 20

Trends in biotechnology and chemical patent practice. Published/Created: New York, N.Y.: Practising Law Institute, Related Authors: Practising Law Institute. Description: v.: ill.; 22 cm. Current Frequency: Annual ISSN: 1049-4294 Notes: Prepared for distribution at the Trends in biotechnology and chemical patent practice program. Description based on: 1989. SERBIB/SERLOC merged record Subjects: Biotechnology--United States--Patents. Chemicals--United States--Patents. Series: Patent, copyright, trademark, and literary property course handbook series Variant Series: Patents, copyrights, trademarks, and literary property course handbook series LC Classification: KF3133.B56 T74 Dewey Class No.: 346.7304/86 347.306486 20

Turney, Jon. Frankenstein's footsteps: science, genetics and popular culture / Jon Turney. Published/Created: New Haven: Yale University Press, 1998. Description: ix, 276 p., [16] p. of plates: iil.; 24 cm. ISBN: 0300074174 Notes: Includes bibliographical references (p. 249-270) and index. Subjects: Science--Social aspects. Science--Public opinion. Biotechnology--Social aspects. Genetic engineering--Social aspects. Science in literature. LC Classification: Q175.5 .T87 1998 Dewey Class No.: 303.48/3 21

United States. Congress. House. Committee on the Judiciary. Subcommittee on Courts, Intellectual Property, and the Administration of Justice. Biotechnology patent protection: hearing before the Subcommittee on Courts, Intellectual Property, and the Administration of Justice of the Committee on the Judiciary, House of Representatives, One Hundred First Congress, second session, on H.R. 3957 and H.R. 5664 ... September 25, 1990. Published/Created: Washington: U.S. G.P.O.: For sale by the Supt. of Docs. Congressional Sales Office, U.S. G.P.O., 1991. Description: v, 290 p.: ill.; 24 cm. Notes: Distributed to some depository libraries in microfiche. Shipping list no.: 91-184-P. "Serial no. 122." Item 1020-A, 1020-B (MF) Includes bibliographical references. Subjects: Biotechnology--United States--Patents. Biotechnology industries--Law and legislation--United States. LC Classification: KF27 .J857 1990l

United States. Congress. House. Committee on the Judiciary. Subcommittee on Intellectual Property and Judicial Administration. Amending Title 35, United States Code, with respect to patents on certain processes: hearing before the Subcommittee on Intellectual Property and Judicial Administration of the Committee on the Judiciary, House of Representatives, One Hundred Third Congress, first session, on H.R. 760 ... June 9, 1993. Published/Created: Washington: U.S. G.P.O.: For sale by the U.S. G.P.O., Supt. of Docs., Congressional Sales Office, 1994. Description: iv, 114 p.; 24 cm. ISBN: 0160444403 Notes: Distributed to some depository libraries in microfiche. Shipping list no.: 94-0212-P. "Serial no. 32." Includes bibliographical references. Subjects: Biotechnology--United States--Patents. Biotechnology industries--Law and legislation--United States. Biotechnology--Technological innovations--Government policy--United States. LC Classification: KF27 .J857 1993j Dewey Class No.: 346.7304/86 347.306486 20

United States. Congress. House. Committee on the Judiciary. Subcommittee on Intellectual Property and Judicial Administration. Biotechnology development and patent law: hearing before the Subcommittee on Intellectual Property and Judicial Administration of the Committee on the Judiciary, House of Representatives, One Hundred Second Congress, first session, November 20, 1991. Published/Created: Washington: U.S. G.P.O.: For sale by the U.S. G.P.O., Supt. of Docs., Congressional Sales Office, 1993. Description: iii, 111 p.: ill.; 24 cm. ISBN: 0160406803 Notes: Distributed to some depository libraries in microfiche. Shipping list no: 93-0237-P. "Serial no. 98." Includes bibliographical references. Subjects: Biotechnology--United States--Patents. Biotechnology industries--Government policy--United States. Biotechnology--Technological innovations--Government policy--United States. LC

Classification: KF27 .J857 1991u

United States. Congress. House. Committee on the Judiciary. Subcommittee on Intellectual Property and Judicial Administration. Biotechnology Patent Protection Act of 1991: hearing before the Subcommittee on Intellectual Property and Judicial Administration of the Committee on the Judiciary, House of Representatives, One Hundred Second Congress, first session, on H.R. 1417 ... November 21, 1991. Published/Created: Washington: U.S. G.P.O.: For sale by the U.S. G.P.O., Supt. of Docs., Congressional Sales Office, 1993. Description: iv, 291 p.; 24 cm. ISBN: 016040715X Notes: Distributed to some depository libraries in microfiche. Shipping list no.: 93-0240-P. "Serial no. 101." Includes bibliographical references. Subjects: Biotechnology--United States--Patents. Biotechnology industries--Law and legislation--United States. Biotechnology--Technological innovations--Government policy--United States. LC Classification: KF27 .J857 1991r Dewey Class No.: 346.7304/86 347.306486 20

United States. Congress. Joint Economic Committee. Biotechnology summit: putting a human face on biotechnology: hearing before the Joint Economic Committee, Congress of the United States, One Hundred Sixth Congress, first session, September 29, 1999. Published/Created: Washington: U.S. G.P.O.: For sale by the U.S. G.P.O., Supt. of Docs., Congressional Sales Office, 2000. Description: v, 259 p.: ill.; 23 cm. ISBN: 0160612853 Notes: Distributed to some depository libraries in microfiche. Shipping list no.: 2001-0042-P. Includes bibliographical

references. Subjects: Biotechnology--
Economic aspects--United States.
Pharmaceutical biotechnology--
Economic aspects--United States.
Series: United States. Congress. Senate.
S. hrg.; 106-677. Variant Series: S. hrg.;
106-677 LC Classification: KF25 .E2+

United States. Congress. Senate. Committee
on Commerce, Science, and
Transportation. Subcommittee on
Science, Technology, and Space.
Competitiveness of the U.S.
biotechnology industry: hearing before
the Subcommittee on Science,
Technology, and Space of the
Committee on Commerce, Science, and
Transportation, United States Senate,
One Hundred Third Congress, second
session, March 23, 1994.
Published/Created: Washington: U.S.
G.P.O.: For sale by the U.S. G.P.O.,
Supt. of Docs., Congressional Sales
Office, 1994. Related Titles:
Competitiveness of the US
biotechnology industry. Description: iii,
75 p.: ill.; 24 cm. ISBN: 0160445493
Notes: Distributed to some depository
libraries in microfiche. Shipping list no.:
94-0231-P. Includes bibliographical
references. Subjects: Biotechnology
industries--United States.
Biotechnology industries--
Competitions--United States.
Biotechnology--Research--Economic
aspects--United States. Series: United
States. Congress. Senate. S. hrg.; 103-
593. Variant Series: S. hrg.; 103-593 LC
Classification: KF26 .C697 1994

United States. Congress. Senate. Committee
on Foreign Relations. The Convention
on Biological Diversity (Treaty doc.
103-20): hearing before the Committee
on Foreign Relations, United States
Senate, One Hundred Third Congress,
second session, April 12, 1994.
Published/Created: Washington: U.S.
G.P.O.: For sale by the U.S. G.P.O.,
Supt. of Docs., Congressional Sales
Office, 1994. Description: iii, 72: ill.; 24
cm. ISBN: 016044814X Notes:
Distributed to some depository libraries
in microfiche. Shipping list no.: 94-
0283-P. Includes bibliographical
references. Subjects: Convention on
Biological Diversity (1992) Biological
diversity conservation--Law and
legislation. Renewable natural
resources--Law and legislation.
Biotechnology industries--Law and
legislation. Sustainable development--
Law and legislation. Series: United
States. Congress. Senate. S. hrg.; 103-
684. Variant Series: S. hrg.; 103-684 LC
Classification: KF26 .F6 1994d

United States. Congress. Senate. Committee
on Foreign Relations. Subcommittee on
International Economic Policy, Export
and Trade Promotion. The role of
biotechnology in combating poverty and
hunger in developing countries: hearing
before the Subcommittee on
International Economic Policy, Export
and Trade Promotion of the Committee
on Foreign Relations, United States
Senate, One Hundred Sixth Congress,
second session, July 12, 2000.
Published/Created: Washington: U.S.
G.P.O., 2001. Description: iii, 74 p.; 23
cm. Notes: Distributed to some
depository libraries in microfiche.
Additional Form Avail.: Also available
via Internet from the GPO Access web
site. Addresses as of 03/09/01:
http://frwebgate.access.gpo.gov/cgi-
bin/getdoc.cgi?dbnam
e=106senatehearings&docid=f:68041.w
ais (text version);
http://frwebgate.access.gpo.gov/cgi-
bin/getdoc.cgi?dbnam

e=106senatehearings&docid=f:68041.p df. Subjects: Agricultural biotechnology--Developing countries. Series: United States. Congress. Senate. S. hrg.; 106-766. Variant Series: S. hrg.; 106-766

United States. Congress. Senate. Committee on the Judiciary. Subcommittee on Patents, Copyrights, and Trademarks. Biotechnology Patent Protection Act of 1991: hearing before the Subcommittee on Patents, Copyrights, and Trademarks of the Committee on the Judiciary, United States Senate, One Hundred Second Congress, first session on S. 654 ... June 12, 1991. Published/Created: Washington: U.S. G.P.O.: For sale by the Supt. of Docs., U.S. G.P.O., 1992. Description: 144 p.; 28 cm. ISBN: 0160375622 Notes: "Printed for the use of the Committee on the Judiciary." "Serial no. J-102-25." Includes bibliographical references. Subjects: Biotechnology industries--Law and legislation--United States. Biotechnology--United States--Patents. Series: United States. Congress. Senate. S. hrg.; 102-457. Variant Series: S. Hrg.; 102-457 LC Classification: KF26 .J863 1991d

United States. Congress. Senate. Committee on the Judiciary. Subcommittee on Patents, Copyrights, and Trademarks. The genome project: the ethical issues of gene patenting: hearing before the the Subcommittee on Patents, Copyrights, and Trademarks of the Committee on the Judiciary, United States Senate, One Hundred Second Congress, second session, on issues relating to genetic research and biotechnology, focusing on the ethical implications of gene patenting, September 22, 1992. Published/Created: Washington: U.S.

G.P.O.: For sale by the U.S. G.P.O., Supt. of Docs., Congressional Sales Office, 1993. Description: iv, 240 p.: ill.; 24 cm. ISBN: 0160416108 Notes: Distributed to some depository libraries in microfiche. Shipping list no.: 93-0606-P. "Serial no. J-102-83." Includes bibliographical references. Subjects: Genetic engineering--United States. Biotechnology--United States. Ethical problems--United States. Genetics--Research--United States. Series: United States. Congress. Senate. S. hrg.; 102-1134. Variant Series: S. hrg.; 102-1134 LC Classification: KF26 .J863 1992c Dewey Class No.: 174/.9574 20

United States. International Trade Administration. High technology industries--profiles and outlooks. Biotechnology. Published/Created: Washington, D.C.: U.S. Dept. of Commerce, International Trade Administration: For sale by the Supt. of Docs., U.S. G.P.O., 1984. Description: 217 p.; 28 cm. Notes: "July 1984." S/N 003-009-00430-6 Item 231-B-1 Includes bibliographical references. Subjects: Biotechnology industries--Government policy--United States Congresses. Biotechnology industries--United States--Congresses. Biotechnology industries--Congresses. Biotechnology--Congresses. LC Classification: HD9999.B443 U68 1984 Dewey Class No.: 338.4/76606/0973 19

Wegner, Harold C., 1943- Patent law in biotechnology, chemicals & pharmaceuticals / Harold C. Wegner. Edition Information: 2nd ed. Published/Created: New York, N.Y., USA: Stockton, 1994. Description: xxxiv, 1029 p.; 24 cm. ISBN: 1561591084 Notes: Includes index. Subjects: Biotechnology--United States-

-Patents. Chemicals--United States--
Patents. Drugs--United States--Patents.
Patent laws and legislation--United
States. LC Classification: KF3133.B56
W43 1994

Wegner, Harold C., 1943- Patent law in
biotechnology, chemicals &
pharmaceuticals / Harold C. Wegner.
Published/Created: New York, N.Y.,
USA: Stockton Press, 1992.
Description: xviii, 514 p.; 25 cm. ISBN:
1561590487 Notes: Errata slip affixed
to p. [viii]. Includes index. Subjects:
Biotechnology--United States--Patents.
Chemicals--United States--Patents.
Drugs--United States--Patents. Patent
laws and legislation--United States. LC
Classification: KF3133.B56 W43 1992

Dewey Class No.: 346.7304/86
347.306486 20

Wiegele, Thomas C. Biotechnology and
international relations: the political
dimensions / Thomas C. Wiegele.
Published/Created: Gainesville:
University of Florida Press, c1991.
Description: x, 212 p.: ill.; 24 cm.
ISBN: 0813010551 (acid-free paper)
Notes: Includes bibliographical
references (p. 181-200) and indexes.
Subjects: Biotechnology--International
cooperation. Biotechnology--Political
aspects. Biotechnology--Social aspects.
LC Classification: TP248.2 .W54 1991
Dewey Class No.: 338.9/26 20

AUTHOR INDEX

M

N

TITLE INDEX

A

A garden of unearthly delights: bioengineering and the future of food, 71

Actinomycetes in biotechnology, 81

Advanced engineered pesticides, 45

Advanced healing technologies, 23

Advanced instrumentation, data interpretation, and control of biotechnological processes, 1

Advanced medical technology, 23

Advanced research on animal cell technology, 125

Advances in applied biotechnology series., 81

Advances in applied lipid research., 81

Advances in biochemistry and biotechnology in Asia and oceania, 108

Advances in bioprocess technology: industrial/specialty chemicals via biological sources/routes., 81

Advances in biotechnology commercialization, 81

Advances in biotechnology for the manufacture of commodity & specialty chemicals., 82

Advances in biotechnology, 1, 81, 82

Advances in chitin and chitosan, 1

Advances in fisheries technology and biotechnology for increased profitability, 53

Advances in gene technology., 82

Advances in heat and mass transfer in biotechnology, 1996, 82

Advances in heat and mass transfer in biotechnology, 1999, 82

Advances in heat and mass transfer in biotechnology--2001, 83

Advances in nucleic acid and protein analyses, manipulation, and sequencing, 83

Advances in plant biotechnology, 45

Advances in plant cell biochemistry and biotechnology., 84

Advances in protein design: international workshop 1988, 84

AgBiotech news and information., 45

AgriBioScan: the agricultural biotechnology directory, 45

Agricultural & environmental biotechnology abstracts., 46

Agricultural applications of biotechnology, 71

Agricultural bioethics: implications of agricultural biotechnology, 46

Agricultural biotechnology & the public good, 46

Agricultural biotechnology at the crossroads: biological, social & institutional concerns, 46

Agricultural biotechnology in developing countries: towards optimizing the benefits for the poor, 47

Agricultural biotechnology in international development, 47

Agricultural biotechnology in the developing world., 47

Agricultural biotechnology in the ICAR research institutes., 47

Agricultural biotechnology, 46, 47, 48

Agricultural biotechnology: a public conversation about risk, 47

Agricultural biotechnology: country case studies: a decade of development, 48

Agricultural biotechnology: issues and choices: information for decision makers, 48

Agricultural biotechnology: opportunities for international development, 48

Agricultural biotechnology: prospects for the Third World, 48

Agricultural biotechnology: the next, 48

Agricultural commodities as industrial raw materials., 49

Agricultural consolidation antitrust, 77

SUBJECT INDEX

C

D

E